Indoor Photovoltaics

Scrivener Publishing
100 Cummings Center, Suite 541J
Beverly, MA 01915-6106

Publishers at Scrivener
Martin Scrivener (martin@scrivenerpublishing.com)
Phillip Carmical (pcarmical@scrivenerpublishing.com)

Indoor Photovoltaics

Materials, Modeling and Applications

Edited by

Monika Freunek Müller

Scrivener Publishing

WILEY

This edition first published 2020 by John Wiley & Sons, Inc., 111 River Street, Hoboken, NJ 07030, USA and Scrivener Publishing LLC, 100 Cummings Center, Suite 541J, Beverly, MA 01915, USA

For more information about Scrivener publications please visit www.scrivenerpublishing.com.

Wiley Global Headquarters
111 River Street, Hoboken, NJ 07030, USA

For details of our global editorial offices, customer services, and more information about Wiley products visit us at www.wiley.com.

Library of Congress Cataloging-in-Publication Data

ISBN 978-1-119-60559-1

Cover image: Top right: GaAs photovoltaic array integrated on a microsystem, placed on top of a US penny. Photo by Inhee Lee and Eun Seong Moon, University of Michigan
Other images provide by the Editor
Cover design by Russell Richardson

Set in size of 11pt and Minion Pro by Manila Typesetting Company, Makati, Philippines

Printed in the USA

Contents

Preface

Indoor photovoltaics (IPV) has grown in importance over recent years. This can in part be attributed to the creation of the Internet of Things (IoT) and Artificial Intelligence (AI) along with the vast amounts of data being processed in the field, which have been a massive accelerator for this development. Moreover, since energy conservation is being imposed as the national strategy of many countries and is being set as a top priority throughout the world, understanding and promoting IPV as the most promising indoor energy harvesting source is considered by many to be essential these days.

There are many contributing factors as to why this book is so special. First of all, I would like to deeply thank all the authors from around the world who found the time to contribute chapters on top of their world-class research, industrial work, and the launch of companies, projects and actual rockets. The book has exceeded my expectations in many ways. Not only did we manage to cover all topics that are essential for building an IPV device, but this book actually contributes to the research in the field. It has yielded collaborations among the chapter authors and the research community, many of which will hopefully be a seed for long-lasting and growing collaborations. Many chapters contain original research that was performed solely for this book, and some chapters contain work that is being published for the first time for a larger audience. On top of all that, the book was already in steady demand even during its writing.

Finally, I would like thank my publisher, Martin Scrivener, for his great and always helpful support, and Jean Markovich, who did outstanding work in copyediting the chapters. I am most grateful for this opportunity and the joy this work brings.

I hope, you, the reader, will find this book helpful in designing and understanding your IPV system.

Monika Freunek Müller
Bern, Switzerland, August 2020

1

Will Photovoltaics Stay Out of the Shadows?

Joseph A. Paradiso

MIT Media Lab, Cambridge, Massachusetts, USA

Abstract
Advances in ultra-low-power electronics capable of useful applications have caused interest in energy harvesting to balloon in recent years. When needed power drops to the point where harvesting is feasible, however, an embedded battery presents serious competition for many use cases, and the particulars tend to rapidly become niche-driven, as they depend on the ambient energy environment that the system experiences. Photovoltaics have long been a popular energy harvesting option, and recent developments expand their opportunities for low-light indoor applications.

Keywords: Indoor photovoltaics, energy scavenging, low-power electronics, human-powered systems, Internet of Things

1.1 Introduction

I'm going to start this book introduction with a confession. Although I'm known for developing a variety of energy harvesting approaches and implementations dating back a couple of decades now, I've never actually really used any of them. Yes, we and others back then and since [1–4] siphoned parasitic power from various modes of human motion and activity - for example, from footfalls alone, my students and I made compact shoe-mounted piezo harvesters in the late 1990s that sent radio transmissions and shoe-integrated magnetic generators that could power

Email: joep@media.mit.edu

Monika Freunek Müller (ed.) Indoor Photovoltaics: Materials, Modeling and Applications, (1–8)

a music system with loudspeakers [5, 6]. Others since have done more evolved work in siphoning power from the foot [e.g., 7, 8], and even products or online DIY projects have appeared [e.g., 9–11], but nothing has hit even close to the mainstream or for that matter even established a strong niche market. The dream of capturing energy from human motion is an opium to the public that hearkens to the appealing magic of perpetual motion machines since there is an eagerness to see this kind of invention. But when the rubber meets the road, even though ingenious implementations have been designed, nothing has gained much traction.

This is for a variety of reasons. One is physical; living bodies are evolved to conserve their energy—our tendons act as springs while walking to store and release ambulant power, and our vascular system constricts to avoid excessive heat loss. Hence, pulling significant power from moving or walking can become tiring, and large-area thermoelectrics can feel cold. Another is expense and complication. Building harvesters into shoes, where most tappable human energy resides, can get pricey and can interfere with footwear designs that are already finely optimized for their purpose (walking, running, climbing, etc.). The complication involved in large mechanical harvesters also introduces fragility; again, going back to the shoe, such devices always break there well before the shoe is worn out. Then there is the issue of power distribution; without wiring the body or clothing, the energy needs to be used where it is harvested. Although you can perhaps glean a good fraction of a Watt from a shoe, it's hard to use it there; the power needs to be transported to a smart AR headset, for example, where that level of energy would be welcomed (ditto for pretty much any kind of mechanical harvester—you need to tap limb motion or the inertial reaction of a heavy mass in a backpack [12] to get to this power level). Sure, you could charge a battery in your shoe and move it—people have implemented this several times. But even with multiple wearable devices, the convenience of powerline charging them, at least in the wired world, has well outpaced this strategy.

OK, so let me backpedal a little from my leading sentence. There is one energy harvesting technology that I have used in my work routinely, and that's photovoltaics. Energy harvesting, by its nature, is a nichey opportunity. Different kinds of harvesting suit environments with different energy reservoirs subject to the constraints of the application and attached device (e.g., energy levels required/available, the conversion efficiency at that energy, and allowable area/volume/mass for the harvesting technology). If your device is buried inside a wall, in a closed shipping box, or embedded in the human body, for example, there is no light available, and then perhaps, if the expense is warranted, thermal conversion (on a warm/hot steampipe, for example) is used [13], or mechanical harvesting (if there's a lot of vibration, ideally at a higher frequency as in an engine) [14], and RF harvesting of beamed or ambient radio with a backscatter tag can work for very low power applications [15]. Within the body, power is generally proximately inductively coupled from transmitting coils on the skin, but ongoing work looks to harvest small amounts through adaptive RF beamforming [16], vibrational scavenging via inertial reaction and induced strain (including on the heart—an idea with a long legacy [17] but only now nearing implementation [18]) and using the body's own energy transport mechanisms via implantable biofuel cells [19]. But in most places where people spend time, there is light by default, as we need it to see, and here photovoltaics are often a top choice if energy harvesting is mandated.

Solar cells offer a harvesting mechanism that has been applied across many orders of magnitude of scale and plant capacity, from urban power systems to spacecraft to mm-scale "smart dust" [20] stacked IC dies gestating in research labs that glean power from a top silicon layer with a micro embedded solar array that also doubles as a coarse video sensor [21]. Photoelectric conversion technology has developed enormously since Einstein won the Nobel Prize in part for explaining it a century ago, thus playing his role (to his purported later chagrin) in ushering in the dawn of quantum mechanics. They have gone through decades of laborious development, enabled through (and likewise inspiring) advances in materials, power conversion and energy storage. Indeed, they will play a major role in renewable energy, together with safe, economical nuclear power and within a few decades fusion reactors designed either by humans or the AIs that will augment us by then. The market and societal need for large-scale solar power is now clear, and has driven the technology and manufacturing into a variety of performance achievements with different form factors while lowering the cost. Thanks to Elon Musk, one notable player out of many in this field, I can now tile my roof with aesthetically-appealing solar shingles and power my Tesla electric car significantly off-grid, even with the available solar illumination up here in New England. The world is now lacking only in political will to buck economic inertia and transition to an emission-free/renewable energy basis in which the capture and harnessing of incident solar photons will be a major part.

That, of course, is the topic of many other books. But the broad, recent expansion of photovoltaics and associated research has also driven the lower end, making light conversion increasingly feasible for low-power embedded applications. Already as a child growing up in the 1960s, I delighted in buying the cheap selenium solar cells they used to sell at the Radio Shack (the dominant electronics component chain store in the US that enabled generations of fellow DIY-ers before it recently went out of business) and hooking them up to small motors, etc., to see them work with a little bit of daylight. With decades of subsequent improvement, solar cells became cheaper and better—certainly at the microwatt level, if you could afford the surface area for some cm's worth of cell area, they have powered simple LCD calculators with indoor light since the 1980s, and watches have even hidden tiny solar arrays in their face—we were proud back then to show off these versions of common gadgets that never needed batteries replaced. But they stayed nichey, as in those days there wasn't much you could do when constrained to microwatts.

Moving to recent times, the long-anticipated explosion of wireless devices in our living environments has taken hold, and simultaneously the amount of function you can achieve at lower power has expanded. Processing, sensing, and even wireless communication have exploited advances in microelectronics fabrication/design, smart power management, and frameworks/protocols to do much more with fewer ergs, while energy conversion and dynamic optimal load matching has become much more efficient, even at the microwatt level and for a variety of different source impedances. This increased function per microwatt has boosted what can be done with a little bit of exposed PV cell, and applications have followed. Already in the mid-1990s we designed and deployed scores of smart IR beacons for indoor location at the MIT Media Lab that ran off light from proximate florescent bulbs [22]. During the last decade, we have run several of our embedded wireless sensing systems from small PV cells (e.g., our Tidmarsh environmental sensing node powers from a tiny solar cell atop its package; a few hours of diurnal daylight or a bit longer in overcast skies will serve to keep it powered [23]), and for low duty-cycle applications, we have built similar embedded light-powered sensors that work indoors.

This is no longer unusual, and various PV-powered wireless sensors have appeared for years now, some by commercial vendors, like the Schneider HOMES temperature/humidity/CO2 sensor developed by our former collaborators at Schneider Research in Grenoble a decade ago. But nonetheless, although the outdoor market for low-power wireless PV devices is robust (witness, for example, commonplace solar-powered outdoor lighting), few indoor PV-powered sensors have achieved widespread commercial viability. This is mainly due to the fact that as power decreases to the point where energy harvesting becomes an option, an embedded battery can last potentially many years, often approaching either the anticipated device lifetime, the battery's shelf life, or just lasting long enough to make replacement a limited irritation (witness how we tolerated batteries in smoke alarms over the past decades), now often announcing over a wireless link when batteries need to be replaced. Batteries pose the ultimate dilemma to energy harvesting, as they are just so cheap and convenient. Plus, they don't run the risk of failing in darkness, and can deliver brief pulses of high current (e.g., a commodity wireless IoT sensor bought today can draw a couple of amperes from a pair of AA cells for a few seconds when first joining an 802.11 network).

Things are changing now on many fronts, and this situation will evolve. For one, indoor photovoltaics are themselves improving. Although the efficiency of real high-end silicon solar modules in bright sunlight can significantly exceed 20%, recombination and spectral mismatch effects in the low light conditions we find indoors can cut efficiencies at least in half [24]. Indeed, the efficient world of LED and fluorescent lighting that we live in now emits at wavelengths less friendly to canonical silicon, and partly in response, ongoing research in photovoltaics has developed many approaches to converting light efficiently across broader spectral ranges. Exploiting these techniques to exhibit better performance at spectra now commonly exhibited indoors, as well as shifting away from straight silicon to new materials like perovskite, GaAs, and Cd compounds, hint at raising this efficiency by a factor of 2 to 3, albeit introducing some complexity in toxicity, manufacturing, etc.; the following chapters in this book explore this in detail. Similarly, needed power levels are still decreasing, not only for processing and sensing, but also for joining "heavier" mainstream networks like 801.11 or 5G, not to mention continued evolution in their low-power brethren like the now ubiquitous BLE and the multitude of other short-range contenders (e.g., ZigBee, Z-Wave, RF-backscatter, etc.).

The sheer number of wireless devices in the environments we live and work in will also continue increasing; at some point, this will cross a level at which battery replacement is less tolerable, together with a profusion of sensors in hard-to-access areas that still have low but usable levels of light (e.g., for environmental control, material integrity, leakage and structural assessment, safety, etc.). Also, as lighting has become more efficient and networked, and cheap renewable energy drops basic power costs, one possibility would be for areas of dwellings to illuminate for an interval when people aren't present just to charge the sensors; they will tell the smart home or IoT servers when they are running low and need an illumination hit, or just raise the shades to let enough daylight in. Long-range inductive charging is a potential competitor here [25], but efficiency loss with range, orientation, and still unacceptably large EMI are show stoppers here, together with the lack of infrastructure; our homes and offices are amply lit, but don't yet (and probably won't) have large induction coils in the floors and walls (on the other hand, I see Helmholtz-coil-laden closets, draws, and shelves as potential ways to conveniently charge wireless electronics embedded in clothing).

Hence, we're at a crossroads here—the dominance of batteries in low-power sensors and systems may well be close to ending as IPV technology advances and smart environments can coordinate their assets to conserve the energy of distributed wireless embedded systems and even recharge them when needed. Both batteries and photovoltaics are differently weighted in terms of manufacturing energy, toxicity, etc., and it's not clear how much photovoltaics will come out ahead over a device's lifetime, but the race is certainly on, and the incentives will continue.

I'm delighted to see this book arrive now—this is certainly the right time for it, and many of the things I've barely hinted at above are explored much more fully in the chapters ahead. But let's check back in after another decade passes to see what will be energizing the low-power devices scattered throughout our environments—and, in what I currently find most exciting and profoundly important, how the data they produce is used and leveraged. In the context-driven world of the future, every bit will mean something—here's hoping that they all will serve us well!

References

1. Paradiso, J.A. and Starner, T., Energy scavenging for mobile and wireless electronics. *IEEE Pervasive Comput.*, 4, 1, 18–27, 2005.
2. Starner, T. and Paradiso, J.A., Human generated power for mobile electronics, in: *Low-Power Electronics*, vol. 45, C. Piguet (Ed.), pp. 45-1–45-35, CRC Press, Boca Raton, Florida, USA, 2004.
3. Mitcheson, P.D., Yeatman, E.M., Kondala Rao, G., Holmes, A.S., Green, T., Energy harvesting from human and machine motion for wireless electronic devices. *Proc. IEEE*, 96, 9, 1457–1486, 2008.
4. Choi, Y.-M., Gu Lee, M., Jeon, Y., Wearable biomechanical energy harvesting technologies. *Energies*, 10, 1483, 1–17, 2017.
5. Kymissis, J., Kendall, C., Paradiso, J., Gershenfeld, N., Parasitic power harvesting in shoes, in: *Proc. of the Second International Symposium on Wearable Computing*, pp. 132–139, IEEE Computer Society Press, Pittsburgh PA, October 1998.
6. Shenck, N.S. and Paradiso, J.A., Energy scavenging with shoe-mounted piezoelectrics. *IEEE Micro.*, 21, 3, 30–42, 2001.
7. Xu, B. and Yang, L., Force analysis and energy harvesting for innovative multi-functional shoes. *Front. Mater.*, 11, September 2019.

8. Zhao, J. and Zheng, Y., A shoe-embedded piezoelectric energy harvester for wearable sensors. *Sensors*, 14, 7, 12497–12510, 2014.
9. SolePower, Power by Walking - https://www.kickstarter.com/projects/764467377/solepower-power-by-walking-0?ref=discovery&term=power%20generating%20shoes (December 2016) or https://www.thisiswhyimbroke.com/power-generating-shoe-insole/.
10. https://news.wisc.edu/power-walk-footsteps-could-charge-mobile-electronics/, Feb. 8, 2016.
11. https://www.instructables.com/id/Power-Generating-Shoe/, 2009.
12. Wang, J. and Liang, J., Energy Harvesting from Horizontal and Vertical Backpack Movements During Walking, in: *Proc. of the IEEE/ASME International Conference on Advanced Intelligent Mechatronics (AIM)*, Auckland, New Zealand, July 9–12, pp. 798–803, 2018.
13. Park, H., Lee, D., Park, G., Park, S., Khan, S., Kim, J., Kim, W., Energy harvesting using thermoelectricity for IoT (Internet of Things) and E-skin sensor. *J. Phys. Energy*, 1, 042001, 2019.
14. Zu, D. and Beeby, S., Kinetic energy harvesting, in: *Energy Harvesting Systems – Principles and Applications*, T.J. Kaźmierski and S. Beeby (Eds.), pp. 1–77, Springer, 2011.
15. Liu, V., Parks, A., Talla, V., Gollakota, S., Wetherall, D., Smith, J.R., Ambient backscatter: Wireless communication out of thin air, in: *Proc. of ACM SIGCOMM 2013*, pp. 39–50, August 2013.
16. Ma, Y., Luo, Z., Steiger, C., Traverso, G., Adib, F., Enabling deep-tissue networking for miniature medical devices, in: *Proc. of ACM SIGCOMM 2018*, pp. 417–431, August 2018.
17. Ko., W.H., Piezoelectric energy converter for electronic implants, July 15 1969. US Patent 3,456,134.
18. Dagdeviren, C. *et al.*, Conformal piezoelectric energy harvesting and storage from motions of the heart, lung, and diaphragm. *PNAS*, 111, 5, 1927–1932, 2014.
19. Zebda, A., Alcaraz, J.-P., Vadgama, P., Shleev, S., Minteer, S.D., Boucher, F., Cinquin, P., Martin, D.K., Challenges for successful implantation of biofuel cells. *Bioelectrochemistry*, 124, 57–72, 2018.
20. Warneke, B., Last, M., Liebowitz, B., Pister, K.S.J., Smart dust: Communicating with a cubic-millimeter computer. *IEEE Comput.*, 34, 1, 44–51, 2001.
21. Lee, Y. *et al.*, A modular 1 mm^3 die-stacked sensing platform with low power I^2C inter-die communication and multi-modal energy harvesting. *IEEE J. Solid-State Circuits*, 48, 1, 229–243, 2013.
22. Starner, T., Kirsch, D., Assefa, A., The locust swarm: An environmentally-powered networkless location and messaging system, in: *Proc. of the ISWC 1997*, pp. 169–170, 1997.
23. Mayton, B., Dublon, G., Russell, S., Lynch, E.F., Haddad, D.D., Ramasubramanian, V., Duhart, C., Davenport, G., Paradiso, J.A., The

networked sensory landscape: Capturing and experiencing ecological change across scales. *Presence: Teleoperators Virtual Environ.*, 26, 2, 2017.

24. Reich, N., van Sark, W. G. J. H. M., Alsema, E.A., Kan, S.Y., Silvester, S., van der Heide, A., Lof, R.W., Schropp, R.E.I., Weak light performance and spectral response of different solar cell types, in: *Proc. of the 20th EUPVSEC*, pp. 2120–2123, 2005.
25. Kurs, A., Karalis, A., Moffat, R., Joannopoulos, J.D., Fisher, P., Soljačić, M., Wireless power transfer *via* strongly coupled magnetic resonances. *Science*, 317, 5834, 83–86, 2007.

2

Introduction to Micro Energy Harvesting

Monika Freunek (Müller)

BKW AG, Bern, Switzerland

Abstract
Trillions of IoT edge nodes are expected to be installed worldwide in the near future. At the same time, many nations have set ambitious goals in order to reduce the amount of required electric energy. How do we want to power these electronic devices? A promising approach is the use of ambient energy by micro generators. This concept is also known as micro energy harvesting. The following chapter provides an introduction to micro energy harvesting and its history, design challenges, and future work. Different sources of energy and available conversion mechanisms are discussed. Principles and challenges of miniaturized devices are introduced.

Keywords: Micro energy harvesting, micro energy scavenging, internet of things, wireless sensor nodes, edge nodes, energy conversion

2.1 Introduction and History

In 2018, McKinsey identified 104 use cases for IoT applications in 11 sectors with 3000 companies actively working on them [1]. The authors reported a potential market value of $200 billion USD until 2025, and trillions of IoT edge nodes are expected to be installed worldwide in the near future. At the same time, many nations have set ambitious goals in order to reduce the amount of required electric energy.

How do we want to supply edge nodes with energy? A promising approach is the use of ambient energy in order to power electronic systems. This concept is also known as micro energy harvesting. The following chapter introduces micro energy harvesting and its history, the challenges

Email: monika.freunek@gmx.de

Monika Freunek Müller (ed.) Indoor Photovoltaics: Materials, Modeling and Applications, (9–24)

in its design and future work ahead. Different sources of energy and available conversion mechanisms are discussed. The aim of this chapter is to set indoor photovoltaics in the broader frame of Micro Energy Harvesting in general. An introduction to conversion principles, achievable output power, and literature suggestions for further reading are provided for each type of converter. The principles of miniaturized and indoor photovoltaic devices are the core topic of this book with a general discussion of the theoretical foundations in Chapter 3.

2.1.1 Brief History of Electric Generators and Loads

The use of ambient energy in human operated applications is far older than the electric grid. Wind- and watermills have been constructed and operated by humans over thousands of years. Based on the work of Bélidor [2], the first electrodynamic hydro generators were developed and applied for local grids, such as by the US Grand Rapids Electric Light and Power Company in 1880 [3].

The following decade mainly focused on large-scale electrification in both appliances and connections. According to the World Bank, 89% of the world was considered electrified in 2017 [4]. Grids were generally constructed as alternating current (AC) grids. Especially after the introduction of nuclear plants, electric appliances were mostly designed for an abundance of power. With the progress in semiconductor research, transistor elements were invented in 1947 [5]. This enabled the development of electronic products based on semiconductor elements, such as silicon. Silicon with its bandgap of 1.12 eV has an open circuit voltage of around 0.6 V. A serial connection of many silicon elements then yields an operating DC voltage around 3 to 12 V, which can be provided by batteries. Therefore, these devices were and are preferably battery powered.

Both the small scale and the low required power of electronic devices enabled miniaturization of products. In 1997, Chu *et al.* demonstrated a cubic corner reflector below the centimeter scale [6]. This was the beginning of the Smart Dust project [7], which aimed at fabricating distributed wireless sensor nodes in a microscale. With a similar approach, DARPA launched a program in order to developed miniaturized wireless sensor nodes [8].

2.1.2 Forms of Energies and Energy Converters

In order to do work on an object, exergy needs to be available. If a system is to use the exergy of another system, which might be surrounding energy or a converter, the systems have to be coupled. The form of coupling between energies of the same form or between different types of energy are called weak and strong. There are nine types of energies [9], and in principle, all of them can be used for energy harvesting:

- gravitational
- kinetic
- heat
- elastic
- electric
- chemical
- radiation
- nuclear
- mass.

Besides the use of mass and gravitational energy, each form of energy has been used in micro harvestings research so far. Reviews on the principles, the history and the current state-of-the-art in industry and research have been provided by various authors, such as [10–12]. As this book focuses on indoor applications, this chapter discusses the level and usability of these types of energy for indoor applications. Some of the most important references are provided for further reading, and principles and examples of demonstrated systems or commercially available devices are included where possible. As most generators aim at converting energy from ambient energy to electric energy, the converters discussed in this chapter will focus on those.

2.2 Kinetic Energy

Kinetic energy can be found in many places in indoor environments, often in the form of vibrating objects or motion from humans and machines. Besides which, there is often flow from both natural and technical sources, such as air or liquid flow streams within buildings, and from air conditioning plants. Amongst the most classic electric converters using kinetic energy as incoming energy are turbines, and piezoelectric generators in forms of sheets.

Kinetic converters transfer oscillations to a mass. Its movements relative to the supporting structures are coupled back under use of the piezoelectric, inductive or capacitive effect onto an electric system. There are various geometries and designs of converters. Reviews, modeling and research work include the work of Mitcheson *et al.*, Roundy *et al.* or Safaei *et al.* [11–14]. Kinetic converters are already available on a commercial scale, with examples being Kinetron or ReVibe [15, 16]. These converters mostly cover industrial applications, including rail and helicopter flight assistance.

The following subsections introduce simple general models and available power densities for some of the indoor applications.

2.2.1 Oscillating Solid Objects

Most kinetic converters for vibrations of solid objects are designed for resonance, which means they achieve their maximum electric power at a certain frequency f for an oscillating mass m. The maximum power P_{max} can be approximated following Roundy *et al.* [17]

$$P_{max} = \frac{\zeta_e}{4(\zeta_m + \zeta_e)} \frac{m}{\omega_f} a^2 \tag{2.1}$$

where ζ_e denotes the electric damping, ζ_m the mechanic damping of the system. For a first approximation, the damping can be modeled to be $\zeta_m = \zeta_e = 0.015$ [17]. In real applications, the modeling of the damping is complex. The total damping ζ follows from $\zeta_m = \zeta_e$, with the damping coefficient $d = \zeta\, 2\omega_f m$, where $\omega_f = 2\pi f$. a is the acceleration of the vibration. The oscillating mass m is damped electrically and mechanically (d). The resonance frequency depends on the feather constant k, where

$$\omega = \frac{k}{m}. \tag{2.2}$$

Figure 2.1 depicts a schematic of the model.

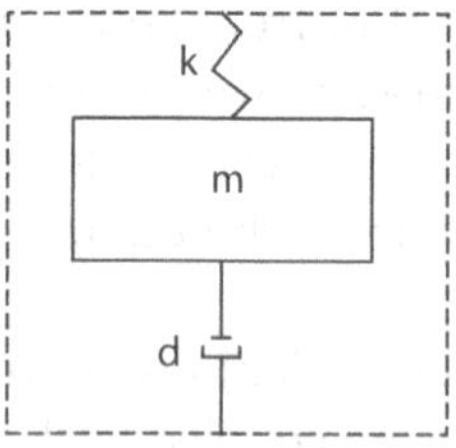

Figure 2.1 Schematic of the basic model of a kinetic energy converter with an oscillating mass m, a feather constant k, and a damping coefficient d.

Following Eq. 2.2, both maximizing the mass m and minimizing the damping d are required for an optimization of the electric power yield. This is the reason why many converters of this type are very heavy. From the models, it also follows that for a constant amplitude, the electric yield increases with decreasing frequencies.

The acceleration of the human lower leg has been measured to be 10 m/s^2 [18], where the frequency f of a human walking is around 2 Hz. For a mass $m = 0{,}2$ kg, and $\zeta_m = \zeta_e = 0.015$ an electric power of 6.64 W could be harvested. Decreasing the mass to a more comfortable $m = 0.01$ kg yields 331 mW. This would already be enough to power small transceivers with low frequencies of use, such as for paging applications in hospitals. As outlined in the introduction of this book by Joe Paradiso, those applications are more of research interest due to their low comfort for the user.

Exemplary demonstrated and commercial systems are outlined in the following sections.

2.2.1.1 *Human Motion*

EnOcean uses kinetic converters in order to power their batteryless radio switches [19]. Kinetron offers weight, rotational and linear based converters with a broad application range including watches, water industry, sanitary applications and similar applications [15].

The PowerWalk system by Bionic Power aims at avoiding soldiers having to carry batteries on their mission by using kinetic converters, with a targeted harvest of 10–12 W [20, 21]. Future applications are aimed at forestry, mining and consumers.

Fu *et al.* demonstrated the use of airflows in micro turbine implemented in a shoe with a power of 6 mW [22]. Many researchers demonstrated the use of piezoelectric converters. A review has been provided by Sun *et al.* [23].

2.2.1.2 *Vibrations*

The commercially available converters for higher frequencies obtain their maximum power at their resonance frequencies, usually 50–60 Hz, and 100–120 Hz, respectively. The bandwidth typically ranges between 1–2 Hz. It follows that the stability of the excitation frequency is their critical parameter. An approach is to order generators that are tuned to application frequencies.

ReVibe Energy sells converters for frequencies from 20–200 Hz with power output from 500 mW to a few W [16]. Figure 2.2 depicts an exemplary generator with equipment from ReVibe Energy.

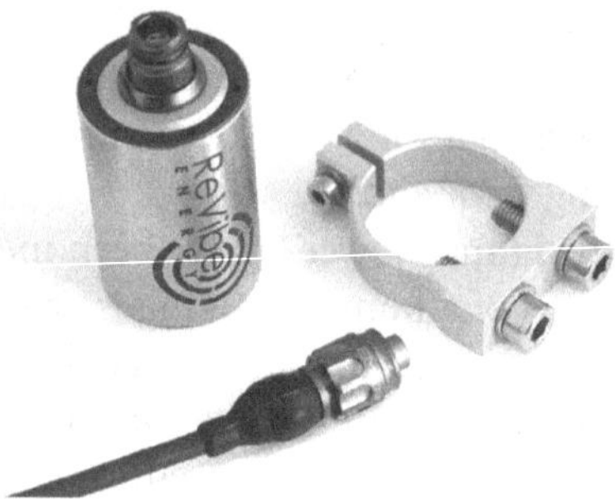

Figure 2.2 ReVibe Energy kinetic converter. (Photo reproduced by permission of ReVibe Energy, 2019).

Non-resonant vibrational harvesters have been a research topic since the beginning of micro energy harvesting. One approach is to tune the electronics [24], the other is to adjust the springs and the length of the resonating mass [25]. Nanoscale generators for implementation in clothes, for example, have been developed by the Georgia Institute of Technology with a corresponding power density of 10 mW/cm^2 [26].

Piezo generators and engineering services for customized generators are offered by different companies, such as Piezo.com [27]. Piezoelectric generators mounted on the wings of beetles yielded 11.5 and 7.5 μW during flight in prototypes from the University of Michigan [28]. The recent state-of-the-art in piezoelectric harvesting has been published by Safaei and coworkers [29].

As in most energy harvesting generators, characterization standards with reproducible test conditions are missing, including test temperature and test frequencies. However, regarding their industrial stage, kinetic converters have achieved a mature state.

2.2.1.3 Flow of Gas and Fluids

The theoretical limit of usable power P_{max} of a gas flow from an electric turbine is defined by the conversion coefficient c_p following the Betz limit [30]

$$P_{max} = \frac{1}{2}\rho A v^3 c_p \tag{2.3}$$

where A denotes the flow area, ρ the density of the flow medium and v its speed. c_p is defined by the ratio of the flow velocity $\vec{v}_1$ before the rotor of the turbine to the flow velocity $\vec{v}_2$ after the rotor

$$x = \frac{\vec{v}_1}{\vec{v}_2}, \tag{2.4}$$

with $c_p(x)$ defined as

$$\frac{P_{out}}{P} = \frac{1}{2}(1+x)(1-x^2) \tag{2.5}$$

and the rotor speed v as

$$v = \frac{v_1}{v_2}2 \tag{2.6}$$

The function has a maximum with x = 1/3 with power coefficient von 16/27. This equals 59.3% of the incoming kinetic power of the theoretical limit of the electrically usable power. The Betz limit covers flow losses due to the design of the rotor blades. Additional mechanical losses are caused by fraction, the inertia of the bearing, and the converter behind the rotor, typically electromagnetic. So far, large wind turbines demonstrated a conversion efficiency of about 40%. New types of design might exceed those limits. However, microturbines are expected to reach a range of 10% due to nonlinear effects and the small ratio of converter to loss surface [13, 31].

The flow speed in HVAC plants can reach up to 6 m/s. For an air density $\rho = 1.2$ kgcm^{-3}, an efficiency of 10% and an area of 1 cm^2, an electric power of 1.3 mW could be achieved. The current European low-pressure systems reach up to 3 m/s, with a resulting power output of approximately 0.2 mW.

Although their maximum power limit is significantly below the limits of large gas flow turbines, indoor air flow of all sources can be a very valuable power source with achievable power in the milliwatt scale. This is sufficient to power most IoT applications.

2.2.1.4 *Acoustic Vibrations*

Acoustic vibrations in indoor environments are available from humans, machines and outdoor vibrations, including traffic. Their harvest is one of the youngest disciplines in Micro Energy Harvesting with applications ranging from typical IoT systems, noise reduction to espionage and hacking. Typical designs are based on, but not limited to, the use of electromagnetic and piezoelectric converters or a combination of both. Yuan *et al.* provided a recent review of the field [32].

Most converters achieve a power in the range of a few microwatts with converters reaching the lower milliwatt scale as the exception.

2.2.1.5 *Elastic Energy*

Elastic energy has had a long tradition in human application, long before electricity was discovered. Watches and clocks have been built based on winding springs that stored energy in its elastic form, which then actuated the pendula and the mechanics. The first turntables and many other devices used the same principles. A common example of an elastic micro energy harvester is the Seiko Kinetic watch.

2.3 Thermoelectric Conversion

Thermal energy can be converted to electric energy with thermoelectric generators (TEGs). In their typical design, TEGs consist of semiconductor materials. Thermocouples, that is pairs of p- and n-type legs, are connected in parallel thermally and in series electrically. The main physical effects in TEGs are the Seebeck, Thompson and Peltier effect, respectively, and Joule heating.

A temperature difference ΔT_g in such a thermocouple generates an electric potential. Following the Seebeck effect, the resulting output voltage U for N pairs of thermocouples is

$$U = N\alpha_s \Delta T_g, \tag{2.7}$$

where α_s denotes the Seebeck coefficient in [V/K] of the thermocouple.

The ohmic resistance causes thermal losses of an electric current I in a conductor of a resistance R, also known as Joule heating P_J

$$P_J = I^2R. \tag{2.8}$$

In a heterogeneous material consisting of materials A and B, an electric flow at the material junction yields a heat flow. This is described by the Peltier effect, where π_{AB} is the Peltier coefficient of the material in [V], with

$$\Pi_{AB} = T\,(\alpha_A - \alpha_B). \tag{2.9}$$

In a homogeneous material, the emission and absorption of heat due to temperature difference in a conductor is represented by the Thomson effect. The Thomson coefficient in [V/K] is

$$\tau_{Th} = T\frac{d\alpha}{dT}. \tag{2.10}$$

The influence of these effects on the TEG performance depends on the size, design and the application environment. Freunek *et al.* have investigated the detailed modeling of the performance of TEG [33].

Any temperature difference of an environment couples into the TEG with a thermal resistance K_g. K_g is optimized when it equals the sum of the coupling resistance of the hot and the cold T_h and T_c side of the TEG, respectively, where

$$K_g = K_c + K_h. \tag{2.11}$$

The maximum achievable temperature difference decreases with increasing miniaturization. The heat sink needs to be dimensioned in a sufficient size, to make a temperature difference usable for a thermoelectric conversion. The size of the heat sink therefore is mostly the limit to miniaturization in TEG systems.

The maximum power output of a TEG P_{out} can be approximated to [33]

$$P_{out}^{max} \approx \frac{\Delta T^2}{4T_k(K_k + K_h)} * \frac{ZT_0}{(1+\sqrt{1+ZT_0})^2}. \tag{2.12}$$

The thermoelectric figure of merit Z is device specific. Figure 2.3 provides a schematic of a TEG.

The output voltage of TEG typically ranges around a few 100 mV. Therefore, DC/DC converters are required.

In indoor industrial applications, the heat from machines can be used with the room temperature as cold sink. In these applications, and especially if the system size is not limited, TEG systems can achieve output powers in the milliwatt range. The same is valid in cold areas or seasons if the temperature difference between outside and indoors can be made usable. However, in most consumer and office applications, the main source for a temperature potential is the difference between wall temperatures and the room temperature. In this case, a TEG will yield a few microwatts.

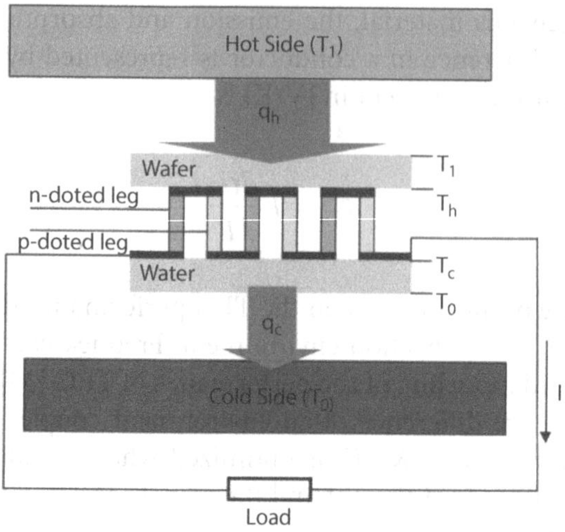

Figure 2.3 Schematic of a TEG with n- and p-type legs, and a hot and cold side.

2.4 Electrochemical Potential

The electrochemical potential from biological sources can be converted to electric energy. Biologic sources have the advantage that the host continuously and actively recharges these power sources by eating and drinking or photosynthesis. Such a host for a biochemical converter can be a plant, an animal or a human.

The first research on the use of biochemical energy, such as adenosine triphosphate (ATP) in the human body, was the ATP powered nanomotor by Cornell University [34]. Voltree Power's bioenergy harvester uses the electrochemical potential resulting from the different pH concentration between soil and a plant [35]. The output voltage was reported to reach 50 to 200 mV for an estimated short circuit current between 0.1 to 1 mA [36]. The induced voltage U for an output voltage U' results from the Nernst equation

$$U = U' - \frac{RT}{nF} \Delta pH \tag{2.13}$$

where R = 8.31447 $\text{Jmol}^{-1}\text{K}^{-1}$ denotes the molecular gas constant, F = 96485,34 C mol^{-1} the Faraday constant, T the temperature, n the number of transmitted electrons and DpH the difference of both potentials.

Other than in some horticultural and agricultural applications, indoor bioelectrochemical potentials are almost solely found in animals and humans. Rasmussen and coworkers demonstrated a biofuel cell with a power of 55 μW, that was implanted in a cockroach [37]. The output voltage was 0.2 V. Such implantable fuel cells can work as an alternative to batteries or cabled solutions with a main application field in medicine. Technical challenges include the low output voltage of around 0.2–0.4 V and the size, stability and nontoxicity of the cathodes. A review of the design, the current state-of-the-art and challenges in biofuel cells has been provided by Zebda *et al.* [38].

2.5 Electromagnetic Transmission

Other than all other forms of energy, electromagnetic emission, including within the optical range, can be sent on purpose for charging applications, as outlined in the introduction of this book by Joe Paradiso. This section focuses on the use of radio frequencies (RF) based on antennas (radio frequency harvesting).

In a simplified model, the efficiency ε_{RF} of these systems is calculated from

$$\varepsilon_{RF} = \frac{P_{DC}}{P} \tag{2.14}$$

The output power thus depends on the emitted power, its distance to the RF harvester, and amount of damping in between. How much power the harvester can receive and convert to usable electric power, depends on the specific antenna design, its damping, and the rectification losses. Recent reviews on this topic have been provided by *Serdijn et al.* and *Cansiz et al.* [39, 40].

The inductive charging of electric toothbrushes is a daily life example of such systems.

2.6 Atomic Batteries

Atomic batteries use the energy of the decay of a radioactive isotope. This energy is converted to electrical energy by different types of generators, such as thermoelectric or optical. Therefore, these systems can be defined as semi-ambient systems. The applied radioisotopes, such as Plutonium-238, are unstable. Their resulting continuous heat emission is used for electric conversion. In the early seventies, some implanted pacemakers were powered by radioisotope batteries based on Plutonium-238 and a TEG.

About ten years later, Lithium-ion batteries superseded atomic pacemakers. Lal and coworkers demonstrated the excitation of a piezoelectric converter with radioisotopes [41]. Currently, atomic batteries have their main application field in space. A review on the topic has been published by Kumar [42].

2.7 Challenges

This chapter focuses on micro energy harvesters for indoor applications other than photovoltaics. Some generators are already available commercially. A current challenge for researchers, product designers and customers is their lack of characterization and test standards. The reproducibility and comparability of test results and data sheets remain difficult, if not impossible, without standardized and representative test conditions.

Toxicity, recycling and lifetime are important topics for future energy harvesting systems, especially for IoT applications. The devices might remain in the field for decades, and some might even be forgotten or left installed after their functional life has ended. This needs to be considered during design. The long lifetime also poses a challenge to the security of IoT edge nodes. Such a system needs to protect and secure its data now and in the future, which might be 15 years ahead. So far, security has been neglected in many IoT and comparable systems. In addition, the visible and invisible interconnection of many edge nodes quickly results in giant networks of an accordingly high attack surface [43].

Finally, a common motivation for the installation of IoT devices is the classical idea of Big Data, that is, gathering as much data as possible in order to realize data analytics applications. However, from a statistical point of view, high quality data from a well-planned sample often yield a better result for these applications than an unlimited amount of data of unknown reliability [43, 44]. Thus, the real value gained from the data obtained by an IoT edge node needs to outweigh its cost to security, privacy, operational effort, toxicity, ecological considerations and so forth.

2.8 Conclusions and Outlook

The use of ambient energy to power low-power electronic devices seems like a natural choice today. This chapter presented the most common types of generators other than photovoltaic converters, their sources of energy, and typical output power levels for indoor applications. From these, kinetic converters including micro turbines and thermoelectric generators are the most mature generators. Commercial products are available on the market, and the output power range is the highest discussed in this chapter. While some of the converters and applications presented in the sections above might seem exotic, and are elaborate in their fabrication or too unreliable for use in commercial products, micro energy harvesting is a valid approach to power IoT edge nodes and applications and a standard in space applications. The appropriate type of generator, its detailed design and the overall design of the IoT edge node depends on the application. Does the system only need to send data at certain events? Does it have to be constantly available or are delays or even outage times allowed? Is the ambient energy easy to predict in its frequency of occurrence and level or is a battery part of the system? A reliable micro energy harvesting system requires a design specific to its application and as a whole system, including its data storage, processing and communication units. An example are kinetic converters, which are tuned to a certain frequency. The need for this application-specific design limits economic mass production effects. Especially when the systems aim for miniaturization, standardized circuit boards that are produced on a mass scale for different product lines cannot be used in micro energy harvesting.

An exception to this, with power levels as high or above the ones presented so far, is indoor photovoltaics, which is introduced in detail in the following chapter.

Acknowledgment

Parts of this work have been translated and revised from the PhD thesis of the author at the Albert Ludwig University of Freiburg, Graduate School of Micro Energy Harvesting DFG GRK 1322-1.

References

1. Chabas, J.M. *et al.*, *New demand, new markets: What edge computing means for hardware companies*, Article, McKinsey & Company, November 2018, https://www.mckinsey.com/industries/technology-media-and-telecommunications/our-insights/new-demand-new-markets-what-edge-computing-means-for-hardware-companies#, accessed 27 January 2020, USA.
2. Bélidor, B.F., *Architecture hydraulique, ou l'art de conduire, d'élever et de ménager les eaux pour les differens besoins de la vie*, 1737–1753, Charles-Antoine Jombert, Paris.
3. Grand Rapids Public Museum, *Grand Rapids*, MI, USA, 2020, https://www.grpmcollections.org/Detail/objects/134698, accessed 27 January.

4. World Bank, https://data.worldbank.org/indicator/EG.ELC.ACCS.ZS.
5. Weiner, C., How the transistor emerged. *IEEE Spectrum*, 10, 1, 24–33, 1973.
6. Chu, P.B. *et al.*, Optical communication link using micromachined corner cuber reflectors, in: *Proc. IEEE The Tenth Annual International Workshop on Micro Electro Mechanical Systems. An Investigation of Micro Structures, Sensors, Actuators, Machines and Robots*, pp. 350–355, Nagoya, Japan, 1997.
7. Kickoff slides from MEMS PI meeting, https://people.eecs.berkeley.edu/~pister/SmartDust/, June 1998, accessed 27 January 2020.
8. Nowak, R. (Ed.), *Proc. of the DARPA Energy Harvesting Program Review*, DARPA, Washington, D.C., USA, April 13–14 2000.
9. Feynman, R.P. *et al.*, *The Feynman Lectures on Physics*, vol. I, California Institute of Technology, USA, 1963, Chapter 4.1 https://www.feynmanlectures.caltech.edu/I_04.html.
10. Paradiso, J. and Starner, T., Energy scavenging for mobile and wireless electronics. *IEEE Perv. Comp.*, 4, 1, 18–27, 2005.
11. Briand, D. *et al.*, Micro energy harvesting, in: *Advanced Micro and Nanosystems*, Wiley, Weinheim, Germany, 2015.
12. Spies, P. *et al.*, *Handbook of Energy Harvesting Power Supplies and Applications*, Pan Stanford Publishing, Boca Raton, USA, 2015.
13. Mitcheson, P.D. *et al.*, Energy harvesting from human and machine motion for wireless electronic devices. *Proc. IEEE*, 96, 9, 1457–1486, 2008.
14. Safaei, M. *et al.*, A review of energy harvesting using piezoelectric materials: State-of-the-art a decade later (2008–2018). *Smart Mat. Struct.*, 28, 113001, 2019.
15. Kinetron, 2020. https://www.kinetron.eu/smart-generators/, accessed 27 January.
16. ReVibe Energy, 2020. https://revibeenergy.com/, accessed 27 January.
17. Roundy, S. *et al.*, *Energy Scavenging for Wireless Sensor Networks: With Special Focus on Vibrations*, Kluwer Academic Publishers, Boston, MA, USA, 2004.
18. Wakeling, J.M. and Nigg, B.M., Modification of soft tissue vibrations in the leg by muscular activity. *J. Appl. Physiol.*, 90, 412–420, 2001.
19. EnOcean, 2020. https://www.enocean.com/en/products/enocean_modules_24ghz/, accessed 27 January.
20. Donelan, J.M. *et al.*, Biomechanical energy harvesting: Generating electricity during walking with minimal user effort. *Science*, 319, 5864, 807–810, 2008.
21. Bionic Power, 2020. https://www.bionic-power.com/#features, accessed 27 January.
22. Fu, H. *et al.*, Energy harvesting from human motion using footstep-induced airflow. *J. Phys. Conf. Ser.*, 660, 1, 2015.
23. Sun, C. *et al.*, On piezoelectric energy harvesting from human motion. *JPEE*, 7, 1, 155–164, 2019.
24. Brenes, A. *et al.*, Tunable unipolar synchronized electric charge extraction strategy for piezoelectric energy harvesting. *J. Int. Mat. Syst. Struct.*, 30, 11, 1629–1638, 2019.

25. Wood, A.R. *et al.*, *EP 2 677 656 A2, Tunable vibration energy harvester and method*, General Electric, Schenectady, USA, 2013.
26. Zhu, G. *et al.*, Flexible high-output nanogenerator based on lateral ZnO nanowire array. *Nano Lett.*, 10, 3151–3155, 2010.
27. PIEZO.COM, https://piezo.com/collections/piezoelectric-energy-harvesters?_=pf&pf_t_quantity=Quantity__1, accessed 27 January 2020.
28. Aktakka, E.E. *et al.*, Energy scavenging from insect flight. *J. Micromech. Microeng.*, 21, 9, 095016, 2011.
29. Safaei, M. *et al.*, A review of energy harvesting using piezoelectric materials: State-of-the-art a decade later (2008–2018). *Smart Mat. Struct.*, 28, 113001, 2019.
30. Betz, A., Das Maximum der theoretisch nutzbaren Ausnutzung des Windes durch Windmotoren. *Zeitschrift für das gesamte Turbinenwesen*, 26, 307–309, 1920.
31. Lissaman, P.B.S., Low-Reynolds-number airfoils. *Ann. Rev. Fluid Mech.*, 15, 223–239, 1983.
32. Yuan, M. *et al.*, Recent developments of acoustic energy harvesting: A review. *Micromachines (Basel)*, 10, 1, 48, 2019.
33. Freunek, M. *et al.*, New physical model for thermoelectric generators. *J. Elec. Mat.*, 38, 1214–1220, 2009.
34. Soong, R.K. *et al.*, Powering an inorganic nanodevice with a biomolecular motor. *Science*, 290, 5496, 1555–1558, 2000.
35. Voltree Power, http://voltreepower.com/bioHarvester.html, accessed 27 January 2020.
36. Love, C.J. *et al.*, Source of sustained voltage difference between the xylem of a potted Ficus benjamina tree and its soil. *PLoS ONE*, 3, 8, e2963, 2008.
37. Rasmussen, M. *et al.*, An implantable biofuel cell for a live insect. *J. Am. Chem. Soc.*, 134, 3, 1458–1460, 2012.
38. Zebda, A. *et al.*, Challenges for successful implantation of biofuel cells. *Bioelectrochem.*, 124, 57–72, 2018.
39. Serdijn, W.A. *et al.*, Chap. 4.2 - Introduction to RF energy harvesting, in: *Wearable Sensors: Fundamentals, Implementation and Applications*, E. Sazonov and M.R. Neuman (Eds.), Academic Press, San Diego, 2014.
40. Cansiz, M. *et al.*, Efficiency in RF energy harvesting systems: A comprehensive review. *Energy*, 74, 292–309, 2019.
41. Lal, A. *et al.*, Pervasive power: A radioisotope-powered piezoelectric generator. *IEEE Perv. Comp.*, 4, 1, 53–61, 2005.
42. Kumar, S., Atomic batteries: Energy from radioactivity. *J Nucl. En. Sci. Power Generat. Technol.*, 5, 1, 2016.
43. Freunek, M., Theory of information for data science systems, BKW AG, Switzerland, unpublished, 2019.
44. Dominitz, J. and Manski, C.F., More data or better data? A statistical decision problem. *Rev. Econ. Stud.*, 84, 4, 1583–1605, 2017.

3

Introduction to Indoor Photovoltaics

Monika Freunek (Müller)

BKW AG, Bern, Switzerland

Abstract

Indoor photovoltaics (IPV) is one of the most promising technologies to power IoT and consumer devices in indoor environments. Theoretical efficiency limits exceeding 50%, an established technology, and power densities among the highest of all indoor energy harvesting technologies available, make IPV attractive and an upcoming technology in an expected trillion-dollar market. However, the typical indoor spectral conditions, the optimization goals and processes during design and manufacturing, and the ideal materials differ significantly from the well-established outdoor applications. Thus, IPV devices require specific engineering and characterization. This chapter introduces IPV with its spectra, irradiance levels and operating conditions. Theoretical and demonstrated efficiencies, their modeling, and characterization methods are discussed. Issues such as total energy balance, recycling and lifetime are outlined. With properly adapted design, IPV might be the enabler of many IoT applications in the near future. This chapter provides the state-of-the-art, challenges and solutions for IPV products, and gives an outlook on future products.

Keywords: Indoor photovoltaics, micro energy harvesting, internet of things, wireless sensor nodes, edge nodes, indoor photovoltaic efficiencies, indoor photovoltaic materials

3.1 Introduction

Indoor photovoltaics (IPV) started in the seventies following the invention of microprocessors [1, 2]. In the following decades, IPV was mostly limited to applications such as kitchen scales, solar calculators, and some

Email: monika.freunek@gmx.de

Monika Freunek Müller (ed.) Indoor Photovoltaics: Materials, Modeling and Applications, (25–38)

consumer fun articles [3, 4]. Those devices were designed far beyond their theoretical potential and often suffered from lack of user acceptance, as users most often had to actively place the device close to a window for a proper functioning. In many applications, it was easier and more reliable to power indoor products with batteries or power cords. The introduction of narrow-banded artificial light and ultra-low power electronics, such as the Phoenix Processor with a power consumption in the picowatt range [5], are the technology enabler for IPV, enhancing their applications far beyond those listed above. Recent studies expected IoT systems to become a multi-trillion-dollar market within the next few years [6]. In many applications, IPV will be the ideal technology to supply the power for the IoT devices.

Indoor photovoltaics presents an ideal application for photovoltaic systems. Due to the narrow-band emission of indoor lighting sources, spectral losses are far beyond the losses from broadband thermal radiation, such as solar. The often mandatory lighting conditions have to meet a certain and defined level, spectral distribution and frequency of occurrence. This knowledge and the predictability of the light source are an ideal case for photovoltaic design. The humidity and temperature operating conditions are mostly stable and not critical to operation and aging. The temperature range is usually around 20–30 °C, which is roughly the standard characterization temperature in photovoltaics. Most loss mechanisms in photovoltaic materials increase with temperature, where indirect band-gap materials, such as silicon, are more affected than direct band gaps such as III-V materials. While outdoor systems under solar radiation heat up to temperatures around 80 °C, temperature effects can be neglected in most IPV applications. Outdoor photovoltaic plants, especially when installed in private applications, often suffer from the expectation of customers to meet the performance of their characterization under IEC [7], which is the standard testing reference for research and industry. Those conditions cannot be met outside a laboratory [8], and in practice, the performance varies with weather, latitude, longitude and air mass composition [9]. Thus, customers sometimes complain about the seemingly poor performance of perfectly operating devices, as the spectral and temperature conditions may deviate significantly from conditions under IEC. Especially for owners of private installations without much technical background, it can be hard to understand this difference, which sometimes lowers the customers' acceptance. This challenge should be much less present in IPV applications, once an indoor characterization standard has been established for IPV.

With an optimized band gap and resulting choice of material, IPV efficiencies exceeding 50% can be achieved theoretically [10].

Besides the spectral conditions, the typical cell size and the available radiation intensity differ from outside applications. While outdoor intensities range around 100–1000 Wm^{-2}, indoor intensities are typically between 0.1–10 Wm^{-2} [11–15]. As a result, the ratio of photoelectric current to electrical loss currents is reduced by orders of magnitude for many loss mechanisms.

The small size in the square centimeter or millimeter scale instead of square meters leads to a stronger influence of miniaturization effects. The loss effects due to shading from module integration and contacting, contact areas and surface areas cannot be neglected as in outdoor applications, but may reach the scale of the photovoltaic current. In real systems, the dominating loss effects depend on the specific material of choice, its thickness and its characteristics, such as doping and so forth. Thus, it is highly recommended that IPV designers first choose their material and then model realistic values for their spectral application conditions based on their choice.

The applicability of efficiency models depends, among other things, on the choice of material. For example, the well-known Shockley-Queisser limit refers to the electrochemical conversion limit of an ideal absorber material where, unlike in real systems, each photon contributes to the photovoltaic current [16, 17]. Indirect band gap semiconductors, such as silicon (Si), have other dominating loss mechanisms than direct band-gap materials such as GaAs, and all of those effects are temperature and intensity dependent. Optimization methods include doping and change of layer thicknesses of the n- and p-layer, respectively. A great introduction into semiconductor modeling in general has been given by Sze and Lee and by Hovel for solar cells in particular [18, 19]. For amorphous Si, these models have to be adjusted according to the material properties and effects [20, 21]. In organic materials, exciton processes have to be modeled [22]. In order to overcome this challenge for IPV, Freunek has recently published a handbook that shows material-dependent and application-specific models for photovoltaic efficiencies [23].

Gemmer, who investigated realistic indoor cell efficiencies for c-Si, a-Si and CIGS with analytical and numerical models, presented the first modeling study for photovoltaic performance under indoor conditions [20, 24]. A current limit to the use of available photovoltaic simulation programs in IPV is their optimization for outdoor applications in their efficiency models and spectra. In addition, in most programs the numerical models have been developed for silicon only and neglect low irradiance or diffuse illumination effects.

For Si, GaAs, and CdTe, Bahrami-Yekta and Tiedje investigated the indoor efficiency limits and the optimization of real devices in indoor conditions in detail [25]. With an absorption layer thickness about two

Table 3.1 Indoor efficiencies modeled by Bahrami-Yekta and Tiedje for different photovoltaic materials (adapted from [25]).

Material	Efficiency FL250 Lux [%]	Efficiency LED [%]
Si	27.0	29.0
GaAs	37.1	40.3
CdTe	40.3	43.3

orders of magnitude below the ones of standard outdoor cells, Si devices can achieve or even outperform their outdoor performance under artificial light. Table 3.1 summarizes their results for three materials. It is important to note that spectra for fluorescent tubes can vary significantly with manufacturer and lamp type.

As Chapter 6 shows similar results to that, the optimization goals for indoor organic photovoltaics (OPV) are contradictory to outdoor applications. For IPV, the influence of the serial resistance in organic devices can be neglected whereas the parallel resistance needs to be maximized.

Customers, product engineers, and researchers require reproducible and retraceable characterization methods in order to compare research results and products. The efforts to establish a standard characterization process has just begun and is outlined in Chapter 5 of this book. Methods to overcome the current gaps in standards are described in Chapters 4 and 5.

The following sections outline general operating conditions, efficiencies and product aspects for the state-of-the-art at the time of writing.

3.2 Indoor Spectra and Efficiencies

Outdoor spectra are broadband spectra resulting from the thermal radiation of the sun. The human eye is sensitive only to a small fraction of it, which has also been characterized as the human visibility function $V(\lambda)$ (see Chapter 4 and 5 for detailed introductions). The users of artificial light in indoor environments are humans, plants, and animals, so their radiation is optimized for their optical range of sensitivity, such as $V(\lambda)$. Broadband radiation sources in indoor environments are halogen lamps and incandescent bulbs, as well as solar radiation filtered through a window. Depending on the place of installation, windows are manufactured with functional coatings of a spectral transmittance $T(\lambda)$, such as heat or sun protection coatings. In general, the filtering function of these windows is optimized to $V(\lambda)$. Thus, the indoor solar spectrum is different to the outdoor spectrum not only in its intensity, but also in its spectral distribution. Figure 3.1 depicts typical indoor spectra. The window spectra result is from the spectral irradiance of the so-called AM 1.5 solar spectrum [26] that is transmitted through a heat and sun protection coating of a window, respectively (see Chapter 4 for a detailed introduction).

As windows in most applications do not face the sky horizontally, the actual irradiance E will result from the incident power P received on an area A in an incident angle α following Eq. (3.1).

$$E = cos\alpha \frac{P}{A} \tag{3.1}$$

In some applications, a sensor might be directly installed on a window facing the outdoors. However, most devices will be installed within a certain distance to the window. As the intensity is inverse to the square of the distance from the emitter, the incident irradiance is additionally lowered following Eq. (3.2)

$$E \propto \frac{1}{d^2}. \tag{3.2}$$

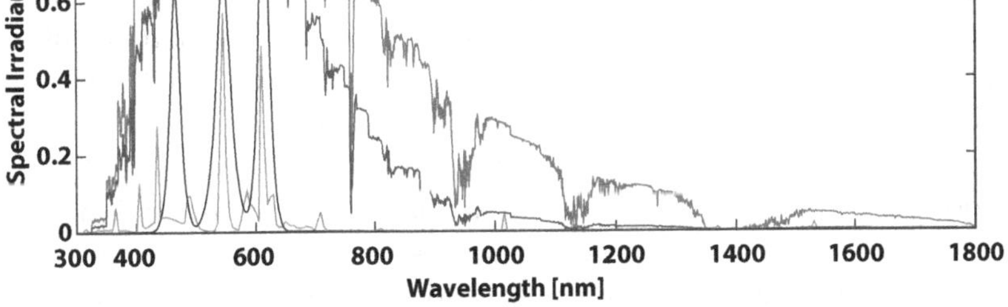

Figure 3.1 Spectra for sunlight through heat and sun protection windows, LED, and warm white fluorescent lamps. The emissions after 1800 nm are negligible for this visualization purpose.

Note that the discussion so far has only included direct radiation. Indirect radiation resulting from reflection or transmission indoors has been neglected. Its effects and modeling are discussed in detail in Chapter 4.

The photovoltaic efficiency limit for a conversion from optical radiation to photoelectric power depends on many parameters. Some of these parameters are spectrally dependent, such as the energy gap or the photon flow, that is the intensity of the radiation. The number of photons $N(\lambda)$ per area A and time t yields the spectral photon flux $\phi_{p\lambda}$

$$\phi_{p,\lambda} = \frac{dN(\lambda)}{Adt}. \tag{3.3}$$

The thermodynamic limit to photovoltaic efficiencies presented by Shockley and Queisser was calculated for black body radiation following Planck's law [16]. It assumes that an ideal single absorber material, where every incident photon is absorbed, yields the so-called external quantum efficiency $EQE = 1$. Then the short circuit current density $j_{sc} \approx j_{ph}$, where the photocurrent density j_{ph} is

$$j_{ph} = q\int_0^\infty \phi_{p,\lambda}(\lambda)EQE(\lambda)d\lambda, \tag{3.4}$$

with the elementary charge $q = 1.602\ 10^{-19}$ As. The Shockley-Queisser limit can be applied numerically to spectral distributions [17, 27]. Figure 3.2 shows the resulting maximum photovoltaic efficiencies for different indoor spectra in comparison to the solar AM 1.5 spectrum.

All indoor radiation is optimized for its visibility by the recipient. For humans, this is the band between 380 nm and 780 nm. Any radiation beyond this narrow band is generally unused and thus lost. The maximum indoor efficiencies therefore are significantly higher than outdoor photovoltaic efficiencies. The radiation filtered in windows outside the visibility spectrum could also be used for photovoltaic conversion.

Resulting from the peak of the human visibility function, the ideal indoor photovoltaic material has a higher band gap than for outdoor applications. Indoor cells optimized to indoor solar radiation have ideal band gaps between 1.4 to 1.6 eV, whilst cells for current artificial light sources have ideal band gaps between 1.8 to 2 eV.

Detailed model and technology discussions for the most interesting indoor materials are provided in the following chapters of this book.

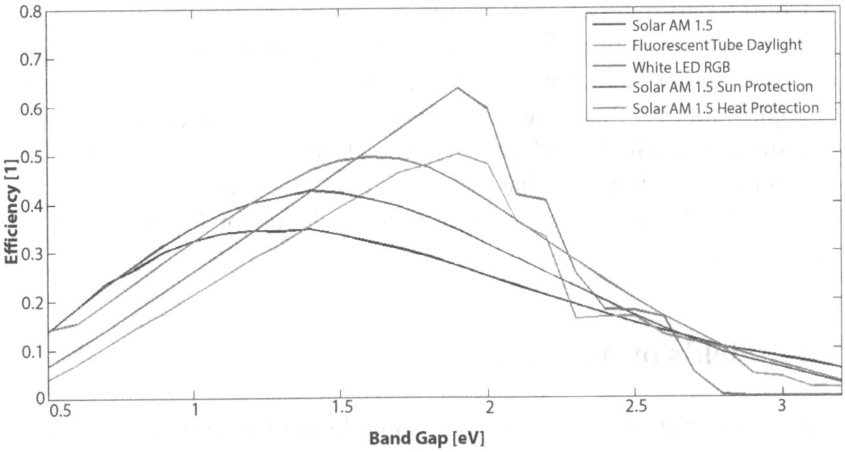

Figure 3.2 Efficiencies for indoor spectra calculated following the Shockley-Queisser limit (adapted from [10]). As a reference, efficiency for outdoor AM 1.5 is given. Due to the adaption of functional window coatings to the human eye, the shift of the ideal band gap to higher values also holds true for indoor solar radiation.

3.3 State of IPV Design, Issues, Approaches

As the irradiance level in indoor surroundings generally is very low, the ratio of active conversion to loss area is very small. Thus, efficiency losses need to be avoided where possible. Therefore, a good indoor design avoids shading caused by module integration, ensures good electric contacts, and aims to avoid recombination losses at surfaces, including edges.

Another issue is a reliable characterization of the device. As standards and equipment are missing, reproducibility and comparability pose a challenge both to designers and customers. This is enforced by the fact that today's most common light sources vary in their spectral distribution and that the available measurement instruments vary in their response to these light sources, too. Furthermore, photometric measurement devices optimized to measure human visibility exclude spectral radiation that is usable for IPV devices. These issues and approaches to solutions are outlined and discussed in Chapters 4 and 5. Müller (Freunek) demonstrated the use of a radiometric reference cell and pyranometers for indoor applications [12, 14]. Today, Enlitech offers an indoor solar cell light simulator, which is capable of measuring both photometric illuminance and radiometric irradiance for low irradiance levels and narrow-band light sources such as LED [28].

Contrary to outdoor applications, where the geographic location and the azimuth might be sufficient for the first stage of planning, in IPV spectra, intensity and contribution of direct irradiance vary from application to application. As these measurements can become too elaborate for the first stage of product development, simulations can be alternatively used. Different research groups have evaluated this for indoor conditions [13, 14, 29, 30]. Chapter 4 introduces the simulation of indoor irradiance conditions.

3.4 Fields of Application

Everywhere humans live and work, some kind of lighting optimized for the human eye is installed. Indoor photovoltaic devices can be operated with it, and often the available light is defined in its spectral range and irradiance following obligatory standards. This eases the design process. Although the classical and well-known IPV products include niche products such as kitchen scales, LED powered lamps, and solar calculators, the potential applications reliably realized with a proper IPV design are as varied as human life. The recent industrial takeoff of the Internet of Things (IoT) with billions of required sensor nodes in the field might drive a wide use of IPV. This section outlines the most common or important application fields for IPV today and in the near future.

3.4.1 Customer and Office Applications

Charging stations for mobile phones are common consumer IPV applications. Further examples include solar calculators, LED powered lamps such as the IKEA Sunnan lamp, USB chargers such as IXYS Green Energy products, and wireless keyboards such as the Logitech K750 [31, 32]. Figure 3.3 shows the Soehnle Easy Solar White kitchen scale.

However, regarding the size of the consumer market, the distribution of products powered by optical radiation is surprisingly low. Reasons for this might be found in spectrally mismatched design, resulting reliability issues, and standard mass production lines that are not optimized for low power consumption.

3.4.2 Ambient Assisted Living and Building Automatization

The concept of ambient assisted living is strongly related to building automation, smart homes, and the IoT. The idea is to support humans in their

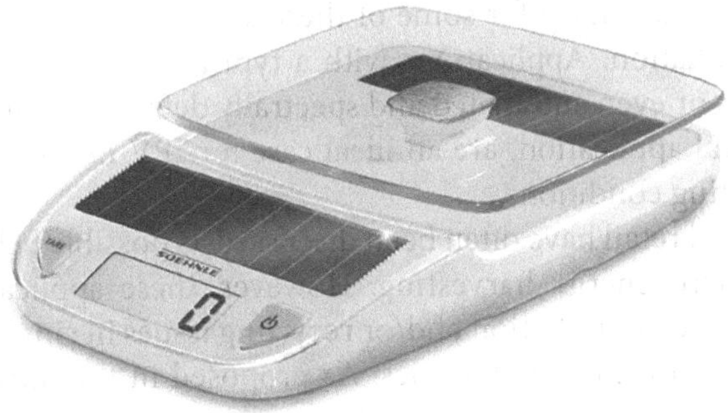

Figure 3.3 Solar powered kitchen scale. (Photo reproduced by permission of Soehnle-Leifheit, 2019).

homes with automated and often customized electronic assistance, often with a focus on the elderly or handicapped individuals. In most applications, reliability is critical. That said, with ambient assisted living still in its early stage of development, this application with its many required sensors has huge potential for IPV; however, so far, the market coverage is very low.

The aim of building automation is to provide comfort and efficiency and to automate energy-saving measures, such as switching lights on and off or regulating room temperatures. Besides financial losses or discomfort, a malfunction of a sensor poses no significant risks. On the other hand, installing automated energy-saving technologies will pay off for most applications very quickly. EnOcean provides different batteryless IPV powered sensors in this application field [33].

3.4.3 Industry, Agriculture, Horticulture, Retail, and Logistics

These applications show some of the highest potential for a significant market growth of IPV, as they are classical applications for IoT and cloud-assisted AI solutions. Typically, the IoT edge node needs to transmit or receive a sensor or actor signal in a regular period or even only under certain circumstances. Therefore, the required average power consumption of such an IoT edge node can range below a few milli- or even microwatts, which can be provided from photovoltaic generators in most cases. There are different optimized products available on the market, including electronics in the case of the TI Bluetooth Low Energy Beacon [34]. IXYS specializes in manufacturing high-efficiency crystalline silicon modules that have been proven to be reliable indoors [10, 35]. The challenge and in some cases the advantage for some of these applications is the frequency of incident radiation. Applications with a typical schedule, such as production hours or even automated and spectrally defined radiation, such as in horticultural application, are an ideal case for IPV design due to their known operating conditions.

Logistics and retail have often been discussed as possible fields of application for micro energy harvesting. However, these applications often suffer from sufficient radiation and/or recycling issues in single-use applications, such as in tracking logistics. As proposed in the introduction, a dedicated radiation power beam might be an approach to some of those applications.

3.4.4 Relation of IPV to Outdoor Applications – Hiking, Emergency Kits

Many small-scale products available on the market were designed for outdoor use or outdoor charging. Among these are emergency kits for hikers, and phone chargers, torches and lamps such as the Little Sun products or the SOLVINDEN table lamp by IKEA [36, 37]. Generally, the module size of these products is a few square centimeters and the material is crystalline or amorphous Si. Although some products use high efficiency cells, they are not optimized for the spectral indoor conditions in their distribution and intensity. Other outdoor-related applications include solar fans for camping or solar power LED lamps. The module sizes for these applications already come close to the range of square meters.

3.5 Degradation and Lifetime Issues

Degradation effects decrease the efficiency over time, thus limit the lifetime among other effects. As most degradation effects are related to energy and intensity levels, indoor conditions are again in favor of the photovoltaic materials. The Stabler-Wronski-Effect in a:Si:H is related to intensity [38], so a:Si:H can expect a longer lifetime indoors than in outdoor operation. As the first generations of outdoor systems start to report lifetimes above three decades, for most materials the photovoltaic module will not be the limiting device for the total device lifetime in IoT edge nodes.

However, so far, few studies have been reported investigating the lifetime for IPV products. In a study from 2013, photodiodes from GaAsP and organic solar cells (P3HT:PCBM) were irradiated by fluorescent light for 3900 hours [39]. This would be approximately 2.3 years of use in an office with 222 working days per year and 8 hours of artificial irradiance per day. The organic cells degraded about 15% from their initial efficiency, the GaAsP about 10%. As these devices might be installed for a long time, degradation issues need to be addressed during design.

A second topic in product design is recycling and toxicity. As IoT devices are small, they might be forgotten in their installations or be integrated into building structure and furniture. Easy removal and recycling and avoidance or handling of toxicity is thus important during early design. Finally, the total cost of ownership and the CO_2 footprint of those devices need to be addressed over the whole production chain and life cycle. Reinders and Apostolou provided an overview of product integrated PV, including IPV and outdoor applications, such as in cars [40]. Chapter 8 introduces current IPV product integration.

3.6 Conclusions and Outlook

The recent takeoff of the IoT could enable the takeoff of IPV. This chapter showed its large potential, and issues to be solved on the way to a safe and secure mass production enabling the full efficiency and reliability of IPV-powered products. The time when we might start to wonder why devices were not powered by IPV might not be so far away.

References

1. Hyatt, G.P., Single chip integrated circuit computer architecture. US Patent 4942516A, 28, Dec. 1970.
2. Faggin, F., Hoff, M., Mazor, S., *Memory system for a multi chip digital computer*, vol. 22, Intel Corp, US Patent 3821715A, Jan. 1973.
3. Conti, J.A., *Electronic postage weighing scale*, vol. 10, Pitney-Bowes Inc, US Patent 4084242A, Nov. 1976.
4. Roen, S.A., *Solar powered portable calculator*, vol. 03, Litton Business Systems Inc, US Patent 4017725A, Jan. 1975.
5. Hanson, S. *et al.*, *IEEE J. Solid-State Circuits.*, 44, 1145–1155, 2009.
6. Atluri, V. *et al.*, The trillion-dollar opportunity for the industrial sector. *McKinsey Digital*, November 2018. Article https://www.mckinsey.com/business-functions/mckinsey-digital/our-insights/the-trillion-dollar-opportunity-for-the-industrial-sector, accessed 2020-01-24.
7. IEC 60904-3:2019, Photovoltaic devices - Part 3: Measurement principles for terrestrial photovoltaic (PV) solar devices with reference spectral irradiance data, 2019.

8. Randall, J.F. and Jacot, J., Is AM1.5 applicable in practice? Modelling eight photovoltaic materials with respect to light intensity and two spectra. *Ren. Energy*, 28, 12, 1851–1864, 2003.
9. Hirata, Y. *et al.*, Variation of output with environmental factors, in: *Photovoltaic Modeling Handbook*, M. Freunek Müller (Ed.), Wiley Scrivener, Hoboken NJ, USA, 2018.
10. Freunek, M. *et al.*, Maximum efficiencies of indoor photovoltaic devices. *IEE J. Photovol.*, 3, 1, 59–64, 2013.
11. Roth, W., *Photovoltaische Energieversorgung für Geräte im kleinen und mittleren Leistungsbereich, report, Fraunhofer-Institut für Solare Energiesysteme*, Freiburg, Germany, 1991.
12. Müller, M. *et al.*, Characterization of indoor photovoltaic devices and light, in: *Proceedings of the 34th IEEE Photovoltaic Specialists Conference*, pp. 738–743, Philadelphia, USA, 2009.
13. Müller, M. *et al.*, *Simulations and measurements for indoor photovoltaic devices, Proceedings of the 24th European Photovoltaic Solar Energy Conference*, Hamburg, Germany, 2009.
14. Müller, M., *Energieautarke Mikrosysteme am Beispiel von Photovoltaik in Gebäuden, PhD Thesis*, Der Andere Verlag, Osnabrück, Germany, 2010.
15. Apostolou, G. *et al.*, Spectral irradiance measurements in a room fit for indoor pv products, *Proceedings of the 27th EUPVSEC*, pp. 4240–4244, Frankfurt, Germany 2012.
16. Queisser, H.J. and Shockley, W., Detailed balance limit of efficiency of p-n junction solar cells. *J. Appl. Phys.*, 32, 510–519, 1961.
17. Trupke, T. and Würfel, P., Fundamental limits of solar energy conversion, in: *Photovoltaic Modeling Handbook*, M. Freunek Müller (Ed.), Wiley Scrivener, Hoboken NJ, USA, 2018.
18. Hovel, H.J., Solar cells, in: *Semiconductors and Semimetals*, vol. 11, R.K. Willardson and A.C. Beer (Eds.), Academic Press, New York, USA, 1975.
19. Sze, S.M. and Lee, M.-K., *Semiconductor Devices: Physics and Technology*, 3rd Ed., Wiley, New York, USA, 2012.
20. Gemmer, C. and Schubert, M.B., Solar cell performance under different illumination conditions. *MRS Online Proc.*, 664, 2001.
21. Shah, A.V. *et al.*, Basic efficiency limits, recent experimental results and novel light-trapping schemes in a-Si:H, μc-Si:H and 'micromorph tandem' solar cells. *J. Non-Cryst. Solids*, 338–340, 639–645, 2004.
22. Thompson, I.R., Modelling of organic photovoltaics, in: *Photovoltaic Modeling Handbook*, M. Freunek Müller (Ed.), Wiley Scrivener, Hoboken NJ, USA, 2018.
23. Freunek Müller, M. (Ed.), *Photovoltaic Modeling Handbook*, Wiley Scrivener, Hoboken NJ, USA, 2018.
24. Gemmer, C.E.M., *Analytische und numerische Untersuchungen von Solarzellen unter wechselnden Beleuchtungsbedingungen, PhD Thesis*, Der Andere Verlag, Osnabrueck, Germany, 2003.

25. Bahrami-Yekta, V. and Tiedje, T., Limiting efficiency of indoor silicon photovoltaic devices. *Opt. Express*, 26, 28238–28248, 2018.
26. NREL, Reference Air Mass 1.5 Spectra, https://www.nrel.gov/grid/solar-resource/spectra-am1.5.html, accessed 28-01-2020.
27. Green, M.A., *Solar Cells: Operating Principles, Technology and System Applications*, Prentice-Hall, Englewood Cliffs, N.J., USA, 1982.
28. https://www.enlitechnology.com/indoor-solar-cell-testing-room-light-simulator.html, accessed 28-01-2020.
29. Reich, N. *et al.*, Weak light performance and spectral response of different solar cell types, in: *20th European Photovoltaic Solar Energy Conference and Exhibition*, vol. 10, pp. 2120–2123, 2005.
30. Apostolou, G. *et al.*, Comparison of the indoor performance of 12 commercial PV products by a simple model. *Energy Sci. Eng.*, 4, 69–85, 2016.
31. https://www.ikea.com/jp/en/catalog/products/00340306/, accessed 28-01-2020.
32. https://www.ixys.com/ProductPortfolio/GreenEnergy.aspx, accessed 28-01-2020.
33. https://www.easyfit-controls.com/de/portfolio/people-activity-counter-epac/, accessed 28-01-2020.
34. https://www.ti.com/lit/ug/tidu235a/tidu235a.pdf, accessed 28-01-2020.
35. http://ixapps.ixys.com/DataSheet/SLMD121H04L_Nov16.pdf, accessed 28-01-2020.
36. https://littlesun.com/product/little-sun-charge/, accessed 28-01-2020.
37. https://www.ikea.com/au/en/catalog/products/70422030/, accessed 28-01-2020.
38. Staebler, D.L. and Wronski, C.R., Reversible conductivity changes in discharge-produced amorphous Si. *Appl. Phys. Lett.*, 31, 4, 292–294, 1977.
39. Freunek Müller, M. and Reindl, L.M., Wirkungsgrade, Charakterisierung und Alterung indoorphotovoltaischer Wandler, in: *Mikrosystemtechnik 2013*, Von Bauelementen zu Systemen, Aachen, Germany, 2013, https://www.vde-verlag.de/proceedings-de/453555132.html.
40. Reinders, A. and Apostolou, G., Product integrated photovoltaics, in: *Photovoltaic Solar Energy: From Fundamentals to Applications*, A. Reinders (Eds.), Wiley, New York, USA, 2017.

4

Modeling Indoor Irradiance

Monika Freunek (Müller)

BKW AG, Bern, Switzerland

Abstract
IPV is one of the most promising and most challenging photovoltaic applications at the same time. Record efficiencies are realistic for this field of applications. However, the intensity and the spectral distribution of available radiation can change within centimeters even for the same room. This chapter discusses methods to measure, calculate and simulate realistic levels of indoor radiation.

Keywords: Indoor irradiance, irradiance modeling, ray tracing, spectral irradiance, indoor photovoltaic, simulation

4.1 Introduction

Only a fraction of the electromagnetic spectrum is visible to the human eye. This range from about 380 nm to 780 nm is called light. Light is available in almost any natural and human surrounding and can be used for photovoltaic generation. In addition, depending on the specific spectral sensitivity of the applied material, parts of ultraviolet and infrared radiation, such as from the sun, can also be used for photovoltaic conversion. In the outdoor area, intensities of solar radiation around 1000 Wm^2 are achievable. Typical values of indoor irradiance and methods of their determination will be discussed in this chapter.

In combination with the generally good predictability of occurrence, the high power densities compared to other forms of energy make optical radiation attractive for the operation of energy-autarkic microsystems in many applications.

Email: monika.freunek@gmx.de

Monika Freunek Müller (ed.) Indoor Photovoltaics: Materials, Modeling and Applications, (39–114)

Optical radiation for indoor applications, such as in building automation or process technologies, is a mixture of direct and diffuse solar radiation filtered by windows, electrical light and heat radiation. In addition to meteorological conditions, the latitude and other location-dependent factors, the amount of usable radiation depends on the distance to a window and the artificial lighting, its type and its frequency of use. A reliable operation of an IPV micro energy harvesting system in an intelligent building depends on both the order of magnitude of typical power densities and the expected spectral distribution. The appropriate modeling boundaries must be evaluated in order to define the optimal choice of generator.

4.2 Fundamentals

The photon formula from Albert Einstein defines the energy E of a photon of frequency f and its wavelength λ, respectively, to be

$$E = hf = h\frac{c_0}{\lambda}, \tag{4.1}$$

where

$$f = \frac{c_0}{\lambda}. \tag{4.2}$$

$h = 6.626*10^{-34}$ Js denotes Planck's constant, and $c_0 = 2.998*10^8$ ms^{-1} the vacuum speed of light, respectively. The photon flow $\Phi_p(\lambda)$ with its unit s^{-1} is the number N of photons per time interval t

$$\Phi_p = \frac{N}{t}. \tag{4.3}$$

The spectral photon flow $\Phi_{p\lambda}$ is often referred to in its distribution over the wave lengths in [s^{-1}nm^{-1}]. The spectral photon flow density $E_{p\lambda}$ in [m^{-2}s^{-1}nm^{-1}] represents the number N of photons per time interval t, area A and per wavelengths interval λ.

For any use of optical radiation, a realistic assessment of the incoming optical power is essential. The incoming power P_{in} is the integral of the number of photons of energy E per area A and time interval t over the incoming wavelength distribution:

$$P_{in} = \int_0^{\infty} \Phi_{p\lambda} \frac{hc_0}{\lambda} d\lambda. \tag{4.4}$$

The irradiance E_e[1] is the ratio of power P per area A or energy E per area A and time t:

$$I = E_e = \frac{P}{A} = \frac{dE}{Adt}. \tag{4.5}$$

Thus, E_e is the energy flow of a radiation on an area in [$Jm^{-2}s^{-1} = Wm^{-2}$]. Taking into account the total spectral distribution of the radiator yields its spectral irradiance $E_{e\lambda}$.

4.2.1 Photometry and Its Impact on IPV

Available data on indoor lighting often refer to photometric illuminance E_v in Lux or lm/m^2. Photometric measures are widely used in the planning of indoor illumination. The illuminance E_v is related to the radiometric spectral irradiance $E_{e\lambda}$ following

$$E_v = K_m \int_{380nm}^{780nm} E_{e\lambda} V(\lambda) d\lambda. \tag{4.6}$$

Photometric quantities have been developed in order to weight an irradiance with the spectral sensitivity function of the human eye. The electromagnetic spectrum which is visible to the human eye ranges from about 380 nm to 780 nm. For the vision in daylight, the visibility function is based on measurements and represented by $V(\lambda)$. The factor K_m denotes the photometric radiation equivalent and is 683 Im/W. Figure 4.1 shows $V(\lambda)$. This figure is also known as the "visibility curve". The sensitivity of the three color receptors of the human eye is different for each color.

E_V is measured with a luxmeter. Generally, a luxmeter consists of a silicon photodiode and a filter which mimics $V(\lambda)$. The spectral sensitivity

[1] E_e is also referred to as intensity I.

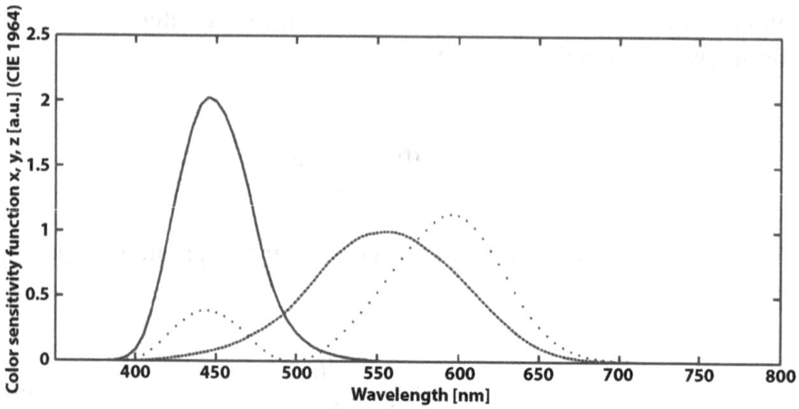

Figure 4.1 Spectral sensitivity function of the human eye for daylight vision with the three color receptors *x* (dots), *y* (dotted), *z* (line), and as resulting total function $V(\lambda)$ (= y).

$S(\lambda)$ of photovoltaic materials, such as germanium or indium gallium arsenide ranges up to wavelength of 1850 nm, whereas crystalline silicon ranges up to 1120 nm. From an IPV point of view, a part of the radiation which could be used for photovoltaic conversion is excluded when measuring with a $V(\lambda)$-filter.

Following equation (4.7), a photometric value can be converted to a radiometric value, if the spectral distribution of the radiation source is known.

The spectral mismatch factor of a luxmeter is an important function for its use. The mismatch factor (m_{mf}) is defined as the quotient of the spectral irradiance $E_{e\lambda}$ of the measured light source and the spectral sensitivity $S(\lambda)$ of the detector to the spectral irradiance $E_{e\lambda cal}$ and the spectral sensitivity $S_{kal}(\lambda)$ of a reference measurement. Ideally, the (m_{mf}) equals 1:

$$m_{mf} = \frac{\int_0^\infty E_{e\lambda cal}(\lambda)d\lambda}{\int_0^\infty E_{e\lambda}(\lambda)S(\lambda)d\lambda}. \tag{4.7}$$

Generally, there is no significant deviation between the spectral sensitivity $S_{kal}(\lambda)$ of the factory standard of a manufacturer and $S(\lambda)$ of the product. Variations can be caused by production tolerances. Quality products document the deviation or calibrate it during production.

For historical reasons, $E_{e\lambda cal}(\lambda)$ consists of a so-called *standard lighting source A* of the Commission Internationale de l'Eclairage (CIE). This is a Planckian radiator with a color temperature of 2856 K (incandescent bulb).

Luxmeters are calibrated to these radiators, as they have been developed in the era of thermal radiators as light sources, such as incandescent light bulbs. These radiation sources emit most of there radiation outside the visible wavelength range. Today they have been widely replaced by emitters of narrow-banded wavelengths, such as LED or fluorescent lamps. Therefore, the influence of the spectral mismatch is much higher for these radiators to the extent that a luxmeter product might not be suitable for an IPV application at all. Figure 4.2 compares the spectrum of a fluorescent tube with daylight spectrum to a *standard lighting source A*. Exemplary measurements of Hildebrandt showed different spectra also for lamps of the same type, such as "white LED" [1].

The measured signal of a luxmeter finally results to

$$E_{Vm} = m_{mf}\, K_m \int_{380nm}^{780nm} E_{e\lambda}(\lambda) V(\lambda)\, d\lambda = \frac{\int_0^\infty E_{e\lambda kal}(\lambda) S_{kal}(\lambda)\, d\lambda}{\int_0^\infty E_{e\lambda}(\lambda) S(\lambda)\, d\lambda} K_m \int_{380nm}^{780nm} E_{e\lambda}(\lambda) V(\lambda)\, d\lambda. \tag{4.8}$$

In most applications, the m_{mf} is unknown, for example due to a missing statement in the technical datasheet or due to an unknown distribution of $E_{e\lambda}$.

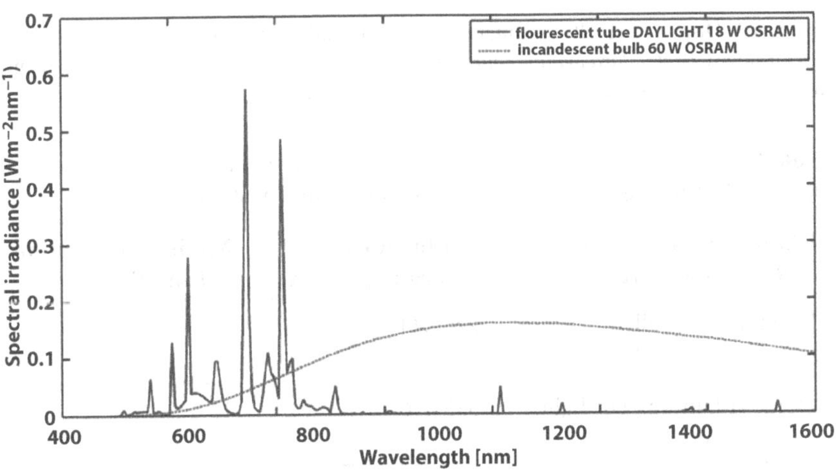

Figure 4.2 Exemplary spectral irradiance of a fluorescent tube and an incandescent bulb.

4.2.2 Comparison Measurements of Different Luxmeter Products and Settings

Disregarding the aspects outlined above, luxmeters often are the first measurement instrument of choice for indoor light measurements or to characterize indoor photovoltaic modules. Figure 4.3 and Table 4.1 depict the measurement results for three exemplary luxmeters of different accuracy and three light sources. The accuracy of the luxmeters refers to a *standard lighting source A*.

The maximum deviation of the measured illuminance E_v of the total mean under a *standard lighting source A* was 4%. For a blue LED the maximum deviation from the total mean exceeded 100% (Table 4.1). For these

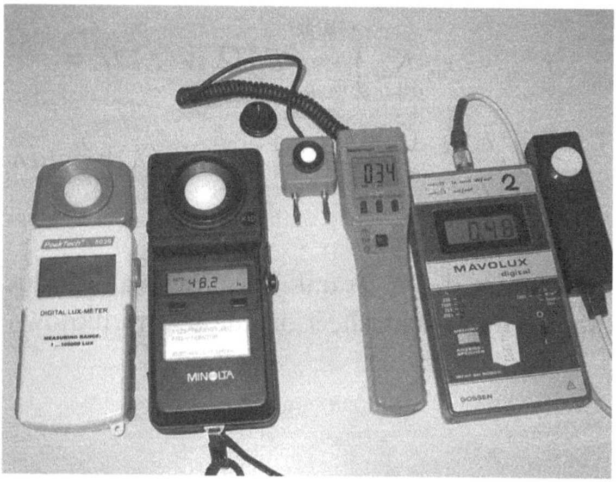

Figure 4.3 Displayed results for different luxmeter products with uncertainties from 2% to 4% under *CIE standard lighting source A*. For the fluorescent light of this measurement, the left luxmeter shows 332 Lux, the middle one 340 Lux.

Table 4.1 Deviation of measured illuminance for different radiation sources when neglecting the m_{mf}: total mean and maximum deviation.

Illuminance E_V in Lux - Radiation source	Arithmetic total mean $\bar{E}_V$ in Lux	Maximum deviation from $\bar{E}_V$ in %
Incandescent bulb CIE Class A OSRAM 40 W	2264.44	3.9
OSRAM Lumilux Daylight 840/18 W	3856.76	16.8
LED Blue	74.97	104.3

results, the m_{mf} has been neglected, as it is a common practice in datasheets of amorphous silicon modules.

4.2.3 Conclusions for Indoor Irradiance Measurements

Currently, there are no standard procedures or commercial devices for the measurement of available indoor irradiance and its spectral distribution. This chapter provides the theory and practical methods in order to obtain reliable results. Chapter 5 of this book outlines current work to obtain international metrological standards for IPV.

At the time of writing, some projects or tests might include the use of a luxmeter. As outlined in the above sections, the m_{mf} and the deviation of the spectral sensitivity of both the detector and the calibrated reference device as well as the calibration spectrum need to be known, and appropriate for the application. If data are published, the type of the luxmeter needs to be part of the documentation in order to enable reproducibility.

The change of preferred lamp types also led to novel types of luxmeters, such as the PCE-LED 20 from PCE instruments [2], which is adapted for white LED light, or the C.A 1110 from Atandra [3], which provides a spectral error compensation for LED and fluorescent tubes. It is important to keep in mind that each type of light source might require its own measurement device. Also, the purpose of all luxmeters is to filter all radiation outside the human visibility function. Thus, the determination of the total available optical power usable for photovoltaic conversion still requires additional steps or different approaches, such as the use of reference cells, reference data, analytical models or simulation. Retractable and reproducible measurement as described in Chapter 5 are essential to any product development, efficiency measurement or a comparison of technologies.

However, during the stages of product development or research projects, the required measurements might be numerous, too expensive or elaborate or training is too complex. The following sections will guide the reader through analytical and numerical solutions.

4.2.4 Available Data on Indoor Irradiance

Most available data on typical spectral distributions and intensities of radiation in buildings refer to photometric use cases. A common example is the standard for the illuminance of working spaces (*EN 12464-1*). This prescribes an illuminance of 500 Lux for office work [4]. In general, there is no standardized spectral distribution, although most office buildings are illuminated by fluorescent tubes with daylight spectrum.

Solar irradiance data are available for various measurement intervals and from different sources, such as the data of more than 2100 weather stations worldwide from the US Department of Energy [5]. More sites are available from the climate simulation program Meteonorm [28].

Scientific studies on typical values of indoor spectral irradiance $E_{e\lambda}$ have been performed by Roth *et al.*, Randall, Roundy *et al.*, and Müller (Freunek) [7–11].

Roundy measured the irradiance as a function of distance from an incandescent lamp and estimated typical indoor irradiance based on standards for office illuminance.

The study of Roth *et al.* investigated irradiance and illuminance for two rooms in Freiburg, Germany. A luxmeter and the short circuit currents of an amorphous silicon module and of a polycrystalline module were used as detectors. The spectral sensitivity $S(\lambda)$, the m_{mf} of the detectors, and the influence of the intensity on the short circuit currents were neglected.

Irradiance and illuminance were measured as a function of distance to the window and of the orientation of the detector area. In a long-term measurement, typical groups of illuminance were classified for different positions in each room. The spectral transmission of the window, the spectrum of the radiation, and the reflectivity coefficients of the rooms were not measured.

For using the results in other rooms, the group recommended taking different variables into account, such as the furniture, the room size, the reflectivity coefficients, the geographical position, user habits or the window size and type [13].

Without further specification, Roth *et al.* listed a presumed global radiation of 600-1000 Wm^{-2} for a clear sky, 200-400 Wm^{-2} for a misty sky and 50-150 Wm^{-2} for a cloudy sky [12].

Randall performed position-depended measurements taking into consideration aspects such as shading from neighboring buildings, spectral transmission from windows with an unstated sensor and basic theoretical investigations [9]. Because of the high variability of the environment, he suggested using ray tracing programs, such as Radiance for further investigations. Roth *et al.* demonstrated an exemplary simulation with *Radiance* for one day in March as part of their study. Müller (Freunek) showed a detailed study with a focus on IPV applications [11], which will be outlined in this chapter.

It is becoming clear from the above section, that the variability of factors such as the presence of users, their habits or installation places of IPV systems remains unsolved with the methods described above. Thus, IPV designers have to perform their own measurements, which are complex, expensive and elaborate, especially for low irradiance. In addition, measurement results are bound to their specific place of measurement. The knowledge of the available radiation is essential for a reliable design of IPV systems. So far, spectral sensitivities of the measurement devices used have not been considered in most studies. Other than in reality, the studied room often consisted of most basic equipment. And finally, despite the significant influence of seasonal differences, no study was conducted for at least one year.

In IPV, the converter area typically ranges in the centimeter scale. In addition to that, most devices are stationary installed. Thus, ray tracing programs for IPV applications require a high physical accuracy. This is in contrast to their more common applications in architecture or film production. Moreover, the simulation needs a meaningful implementation of user behavior. Finally, the simulation needs to be able to perform annual runs in order to simulate the influence of the seasons.

The following sections will cover the use of radiometric solutions and the use of ray tracing programs for IPV applications in detail.

4.3 Radiometric Solutions

This section introduces a radiometric solution based on the ray tracing programs *Radiance* [14] and *DAYSIM* [15] and a user behavior model. This approach enables the evaluation and variation of IPV application scenarios without the need of measurements. Furthermore, the section provides orientation values and shows the most dominant factors influencing the typical irradiance in buildings. The aim is to provide a reliable, cost-efficient and flexible planning of IPV systems.

4.3.1 Structure

The plausibility discussion of a simulation result is as important as setting up and running the simulation itself. It is therefore important to understand which assumptions are made in a simulation model and which results are within a plausible physical range.

The following sections show exemplary analytical calculations for two measurement points. These calculations help to understand the scientific questions and the theory on which the simulations are based. Furthermore, the theoretical limiting range of irradiance in buildings is evaluated. The use of purely analytical models for IPV applications is very limited, mainly due to the sheer number of parameters and their variation. These limitations are discussed in the analytical section.

The following section investigates indoor irradiance based on numerical calculations. The last section of this chapter presents the measurement campaign for the model validation and the evaluation of appropriate radiometric measurement procedures.

4.3.2 Settings of the Studied Rooms

The methods were tested at two offices at the University of Freiburg/Breisgau, Germany. The "South Office" received daylight from a window oriented toward the south east, was sparsely furnished with light materials and mainly used by one person. This office is a good example of sufficient lighting on almost every point in the room. Due to its small amount of furniture, the room is simple to model. Figure 4.4 depicts the room in a hemispherical view. The Figures 4.5 and 4.6 show the room with normal optics from the perspective of someone entering the room, and sitting at the desk, respectively.

The "North Office" received light through a window facing north west. The entrance was 10 m from the windows. Glass sliding doors shielded working places near the windows from the rest of the room, which also blocked incoming daylight (Figure 4.7). The room was about 10 m deep and equipped with many dark furniture pieces, piping, cable routes, and large plants. In total, this room is an example of a scenario with complex application conditions. Figure 4.8 shows a hemispherical view of the room from the entry of the room. Figure 4.9 depicts in normal optics the left side of the room from the entry. The position of the investigated points in the rooms is discussed in the following sections.

Figure 4.4 South office: hemispherical view.

Figure 4.5 South office: view from entry.

Figure 4.6 South office: view from desk.

4.3.3 Investigated Installation Points

IoT devices and autarkic sensor nodes in buildings are likely to be installed on work desks, close to light switches, on the wall or close to a window. Six of such typical installation places were measured, and analytically and numerically modeled. "N" refers to sensor points in the North office, "S" stands for south office.

In addition, lines of virtual sensors with increasing distance to the window and varying orientations were simulated. All simulations included virtual control sensors outside of the rooms. These additional virtual sensors

Figure 4.7 North office: glass doors at the window side.

Figure 4.8 North office: hemispherical view.

Figure 4.9 North office: left side of the room.

were used to control the correct implementation of weather data and as a detector for major simulation errors.

Radiance uses location and orientation vectors for its calculations. Its coordinate system has a fixed orientation to the cardinal directions, as shown in Figure 4.10. The investigated points of installation are summarized as vector description in Table 4.2.

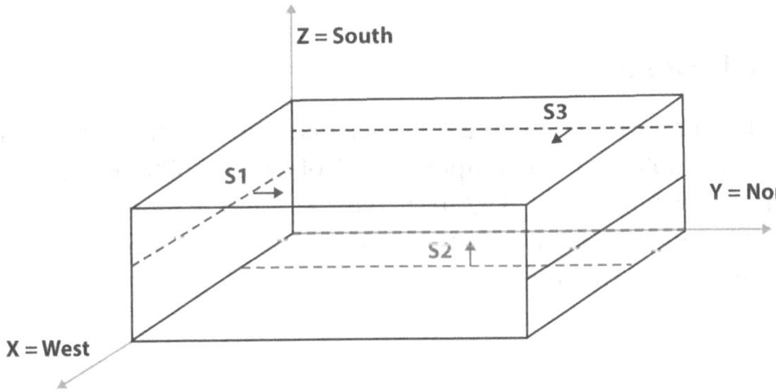

Figure 4.10 Schematic of vectors, the coordinate system and the cardinal directions in Radiance. "S" stands for sensor.

Table 4.2 Notation, orientation, position, and qualitative description of the investigated points of installation.

$\vec{N1}$	$\begin{pmatrix} 0 \\ 1 \\ 0 \end{pmatrix}$	$\begin{pmatrix} 2.55 \\ 0.05 \\ 1.08 \end{pmatrix}$	orientation to window, light switch level, distance to window: 10 m
$\vec{N2}$	$\begin{pmatrix} 0 \\ 0 \\ 1 \end{pmatrix}$	$\begin{pmatrix} 2.80 \\ 5.20 \\ 0.62 \end{pmatrix}$	orientation to artificial lighting, work desk level, distance to window: 4.8 m
$\vec{N3}$	$\begin{pmatrix} 1 \\ 0 \\ 0 \end{pmatrix}$	$\begin{pmatrix} 0.28 \\ 7.20 \\ 1.83 \end{pmatrix}$	orientation into room, head level, distance to window: 3.0 m
$\vec{S1}$	$\begin{pmatrix} 0 \\ -1 \\ 0 \end{pmatrix}$	$\begin{pmatrix} 1.00 \\ 5.29 \\ 1.10 \end{pmatrix}$	orientation to window, light switch level, distance to window: 5.3 m
$\vec{S2}$	$\begin{pmatrix} 0 \\ -1 \\ 0 \end{pmatrix}$	$\begin{pmatrix} 3.41 \\ 0.30 \\ 1.60 \end{pmatrix}$	orientation to window, head level, distance to window: 0.3 m
$\vec{S3}$	$\begin{pmatrix} 1 \\ 0 \\ 0 \end{pmatrix}$	$\begin{pmatrix} 3.69 \\ 0.40 \\ 1.60 \end{pmatrix}$	orientation into room, head level, distance to window: 0.4 m

4.4 Analytical Model

4.4.1 Solar Radiation

In a simplified model, the sun is a spherical black body with isotropic emission. With the solar surface temperature T of around 5780 K, the Stefan-Boltzmann constant σ_{SB} of 5.67*10^{-8} $Wm^{-2}K^{-4}$ and a mean solar radius r_s of 6.96*10^{8} m, the solar radiative power P_{sol} follows from the Stefan-Boltzmann law of radiation:

$$P_{sol} = \sigma s B T^4 4\pi r_s^2 = 3.85 \cdot 10^{26} W. \tag{4.9}$$

With a mean radius of the orbit of the earth around the sun $r_e = 1.496*10^{11}$ m and P_{sol} we can calculate the extraterrestrial solar constant E_0[2]:

$$E_0 \frac{P_{Sol}}{4\pi r_e^2} = 1368.95 \frac{W}{m^2}. \tag{4.10}$$

The solar constant refers to the power received per unit area just outside of the earth atmosphere. The receiving area is normally oriented toward the radiation direction.

The incidence angle of the radiation on its way through the atmosphere is a significant factor of the maximum receivable irradiance on the earth. The photons interact with atmospheric gases and are absorbed and reflected. A direct incidence of the radiation through the atmosphere leads to a so-called Air Mass (AM) Factor 1.0. Standard Test Conditions (STC) following IEC 60904-3 refer to an incident angle of 48.19° with a resulting AM 1.5 [16]. Figure 4.11 compares the spectral irradiance of AM 1.5 with the one of extraterrestrial radiation just before entering the atmosphere of the Earth, AM 0.

Equation 4.11 relates the incident angle to the Air Mass Factor. l_0 is the thickness of the atmosphere,/the way of the radiation through the atmosphere, and Θ_Z the angle against normal to the receiver unit on the earth.

$$AM = \frac{l}{l_0} = \frac{1}{cos\Theta_Z} \tag{4.11}$$

The irradiance E_e also depends on meteorological, seasonal and geographic influences. The following section provides an exemplary calculation of the maximum irradiance on a unit area $A = 1$ m^2 for the investigated building in Freiburg, Germany, and the four most important seasonal points of the meteorological year. One surface is horizontally orientated to the zenith, and two surfaces are at the windows of the investigated offices. From this, the expected values of the direct irradiance for different seasons are obtained.

[2] The Equations 4.9 and 4.10 show the dependence of the extraterrestrial solar constant from the astronomical parameters, such as the solar temperature. The official value from NREL is 1366 Wm^{-2} [17].

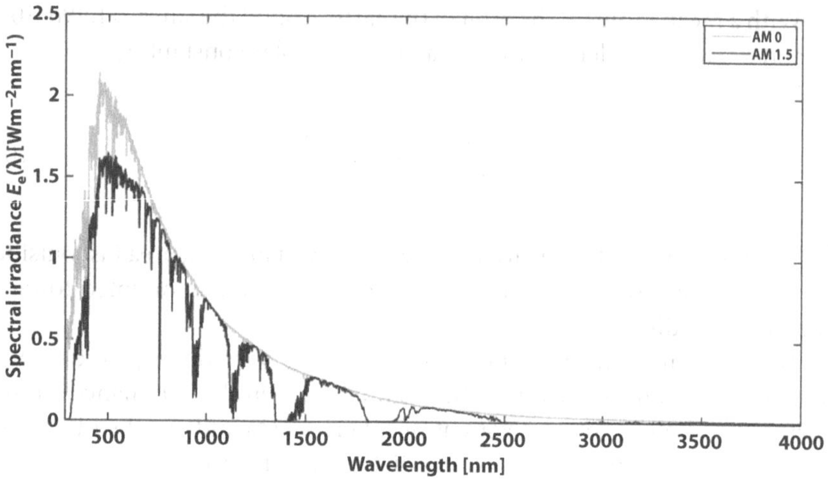

Figure 4.11 Spectral solar irradiance AM 0 and AM 1.5.

For homogeneous irradiance E_e, the incoming radiation flow P_{in} is

$$P_{in} = \int_0^\infty E_{e\lambda} A \, d\lambda. \tag{4.12}$$

We will assume the radiated area A to be $A = 1\ m^2$.[3] The radiation flow P_{in} depends on the incident angle Θ of the solar radiation:

$$P_{in} = E_e A cos\Theta, \tag{4.13}$$

where

$$\begin{aligned} \cos(\Theta) &= \sin(\delta)\sin(\varphi)\cos(s_N) - \sin(\delta)\cos(\varphi)\sin(s_N)\cos(a_Z) \\ &+ \cos(\delta)\cos(\varphi)\cos(s_N)\cos(\omega) + \cos(\delta)\sin(\varphi)\sin(s_N)\,cos(a_Z)\cos(\omega) \\ &+ \cos(\delta)\sin(s_N)\sin(a_Z)\sin(\omega). \end{aligned} \tag{4.14}$$

The latitude φ is + 48.0° (northern hemisphere), the longitude Λ + 7.8° (eastern longitude). The angle s_N between the horizontal surface and the area A is 90° for the windows, and 0° for the horizontal surface, a_Z is the azimuth angle of the receiver area, that is the deviation from an orientation

[3] In radiometric literature, P_{in} can also be found as Φ.

toward the south. ω is the hour angle with morning as + and afternoon as –. δ is the declination, which is the tilt of the axis of the earth relative to the sun.

At the winter solstice, around the 21st of December of the calendar year, the sun is vertical at noontime at 23.5° of southern latitude. The angle Θ_Z then becomes

$$\Theta_Z = (48.0 + 23.5)° = 71.5°. \tag{4.15}$$

This results in an Air Mass Factor *AM* of

$$AM = \frac{1}{cos(71.5)} = 3.15. \tag{4.16}$$

Already this simple calculation gives a powerful impression of the realistic range of indoor irradiance when compared to standard conditions with an AM 1.5. For summer solstice the reference latitude is 23.5° northern latitude with a resulting total angle of 26.5°. The AM for Freiburg then is 1.12. The beginning of spring and autumn on the 21st of March, and the 23rd of September, respectively, refer to the equator. The angle is 48° with a resulting Air Mass Factor of 1.5°. The Air Mass condition from *STC* is therefore approximately met in the discussed test area during spring and autumn.

The irradiance for $E_0 = 1366\ Wm^{-2}$ can be approximated by the Lambert-Beer law.

$$E_e = E_0 e^{-c_E AM}. \tag{4.17}$$

c_E is the extinction coefficient of the atmosphere, which depends on meteorological and geographical conditions. For $c_E = 0.25$ the maximum irradiance through direct solar radiation is between 622 Wm^{-2} (21st of December) and 1032 Wm^{-2} for Freiburg, Germany.

The declination δ can be calculated using the simplified model of Duffie [18]:

$$\delta = 23.45° \sin(360° \frac{284 + N_T}{365}). \tag{4.18}$$

N_T is the day of the year to be modeled, here with $N_T = 1.365$. The hour angle ω varies depending on N_T, the longitude Λ and the reference meridian of the time zone Λ_0. For this, the apparent time t_{ap} needs to be calculated following Duffie [18]

$$t_{ap} = o_t + T_e + (4 * (l_0 - l)) \tag{4.19}$$

where Λ_0 = 15° for the the official time o_t = Central European Time and Λ_0 = 30° for the Central European Summer Time (in 2009 from 29.03. to 25.10.). O_T is calculated in minutes for this chapter. T_e is the result of the solar local time equation in minutes. The time equation incorporates variations of the elliptical orbit of the earth around the sun, which are caused by variations of the rotation of the earth or its orbit [18]. The following calculations are based on the approximation by Spencer [19] where

$$T_e = 229.18(0.0000075 + 0.001868\cos(D)$$
$$-0.032077\sin(D) - 0.014615\cos(2D>) - 0.040849\sin(2D>))$$

and

$$D = N_T(360°/365) \tag{4.20}$$

The hour angle then results to

$$\omega = (12 - t_{ap})15° \tag{4.21}$$

Table 4.3 provides the results following Equations 4.12-4.21 for the four exemplary days within one year and the south-eastern oriented office. Table 4.4 shows the assumed parameters and interim results. t_{ap10h} denotes the point where the indicated official time is reached, here with g_z = 10 h. The calculated maximum value directly on the window glass is 737 Wm^{-2} on the 23rd of September.

The minimum accounts to 350 Wm^{-2} on the 21st of March. Although still idealized in many aspects, these values of maximum achievable irradiance remain significantly below the results from Equation 4.17.

4.4.2 Artifical Lighting

Fluorescent tubes with daylight spectrum are the most common indoor lighting sources in offices. An example of such a lamp is the OSRAM Lumilux 58W/840, which has been installed in the investigated rooms. Its

Table 4.3 Calculated maximum irradiance per unit area for Freiburg, south-east office: a_z = -20°; horizontal *h*: s_N = 0°; window area: s_N = 90°.

Gregorian calendar date	21st of March	21st of June	23rd of September	21st of December
E_{e10hS} in Wm^{-2}	574	932	615	169
E_{e10hSF} in Wm^{-2}	460	350	627	450
E_{el2hSh} in Wm^{-2}	628	894	564	194
E_{el2hSF} in Wm^{-2}	680	476	737	577

Table 4.4 Calculated maximum irradiance per unit area for Freiburg, s = 0° and interim results, apparent time t_{ap} in decimal hours.

Gregorian calendar date	21st of March	21st of June	23rd of September	21st of December
AM	1.50	1.12	1.50	3.15
N_T	80	172	266	355
E_e in Wm^{-2}	939	1032	939	622
δ in °	0.242	23.450	-0.767	-23.450
ω_{10h} in °	24.57	8.20	5.86	22.30
t_{ap10h} in h:min	10.36	11.45	11.61	10.51
ω_{12h} *in* °	-5.43	-21.78	-24.14	-7.70
t_{ap12h} in h:min	12.36	13.45	13.61	12.51

spectrum has been used for the simulations within this chapter. In this section, the received direct irradiance from this artificial light source is calculated for the measurement point *N2* in the north office. The Figures 4.12 and 4.13 depict the measurement point and its rough field of view. Figure 4.14 compares the spectrum of the Lumilux to the terrestrial solar spectrum.

The emission behavior of cylindrical bodies, such as a fluorescent tube, is complex to model (see [21]). This section approximates the specific

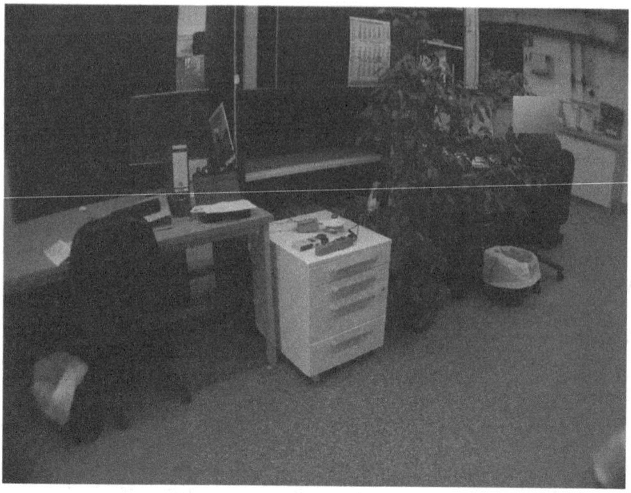

Figure 4.12 Position of the measurement point *N2*.

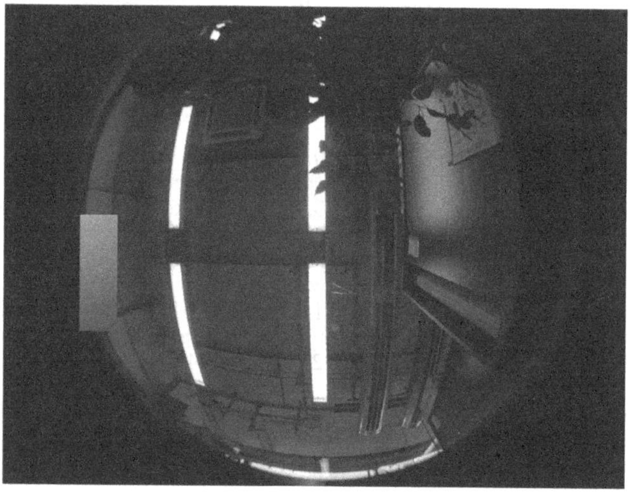

Figure 4.13 Field of view from the measurement point *N2* (hemispherical).

emission of a fluorescent tube (FT) by considering the tube and the receiving area as two concentric cylinders facing each other. With an electric incoming power of 58 W and an efficiency of about 20% for a fluorescent tube the radiation power of the tube P_1 results to 11.6 W.

The radius r_1 of the FT is 1.3 cm with a length of 1.5 m. The distance to the receiving area at *N2* is 2.6 m.

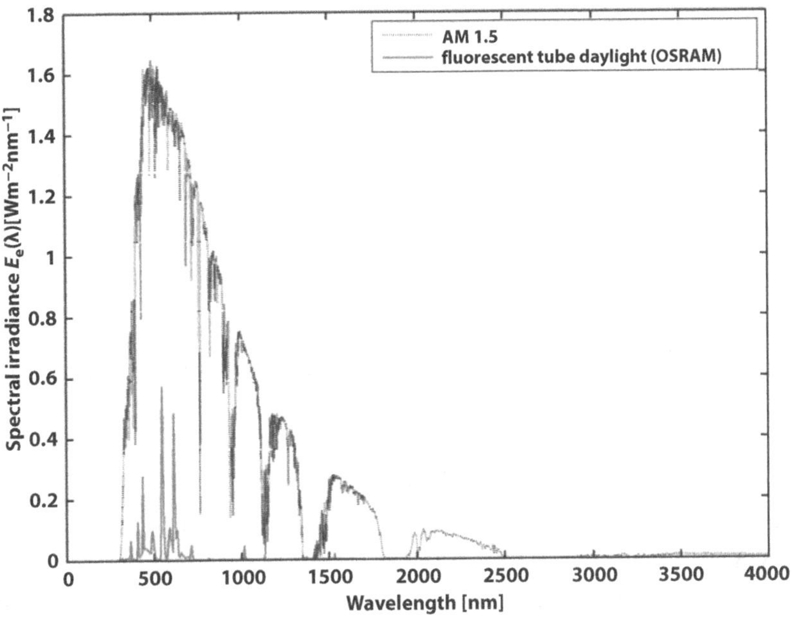

Figure 4.14 Spectral irradiance of a fluorescent tube with daylight spectrum and terrestrial solar radiation AMI.5.

Emitted photons are assumed to move randomly in all directions. Photons in z-direction are reflected at the side surfaces and account completely to I_t. There is only emission in the x- and y-direction, and the Z2-areas do not contribute to the calculation. Figure 4.15 depicts this model. With these simplifications, the irradiance on the measurement point *N2* follows from the lateral surface of the cylinder

$$P_1 = 11.6W \tag{4.22}$$

$$Z = 2\pi r_2 l = 24.38 \tag{4.23}$$

$$E_{N2} = \frac{P_1}{Z} = 0.48\frac{W}{m^2} \tag{4.24}$$

Assuming the emitted radiation is bundled to a quadrant yields a final result of 1.9 Wm^{-2}.

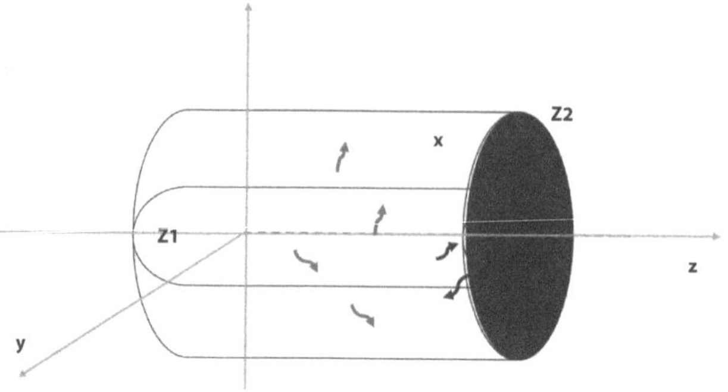

Figure 4.15 Schematic of the assumptions for the calculation of the fluorescent tube from Equation 4.25. All photons in z-direction are reflected.

4.4.3 Interaction with Objects

The interaction of radiation with objects within the interior is most crucial when modeling indoor irradiance, as it affects spectra and intensity. For this section, the interaction is simplified and described based on the coefficients of absorption, reflection and transmission α, τ_t, ρ_r of the radiated object with $1 = \alpha + \tau_t + \rho_r$. An ideal black body absorbs any photon and converts its energy to heat, with $\alpha = 1$ and $\rho_r = 0$. An ideal white body reflects any radiation completely with $\alpha = 0$ and $\rho_r = 1$. Real bodies are in between these extremes.

Transmission

In indoor environments, transmission can be mainly found at the windows. The incoming outdoor irradiance is reduced by the transmission factor. The transmission coefficient can be measured or obtained from the window manufacturer. Depending on the technology and the material, τ_t typically ranges around 0 in the far infrared range and is optimized for permeability in the visible range. For heat protection glass, τ_t is about 0.6 to 0.95 in the visible range. For wavelengths from 300 nm to 2500 nm, a typical transmission coefficient is between 0.4 to 0.6, and is often denoted as g-value. Figure 4.16 shows the spectral transmission of typical isolation and heat transmission windows in comparison to the spectral solar irradiance 1.5 AM. Close to the equator, sun protection windows are used with g-values around 0.3. A similar low g-value can be found for windows with special function coatings.

For a g-value of 0.6 and the maximum outdoor irradiance calculated in Section 4.4.1, the achievable indoor irradiance thus reduces to 210–442 Wm^{-2} due to transmission losses.

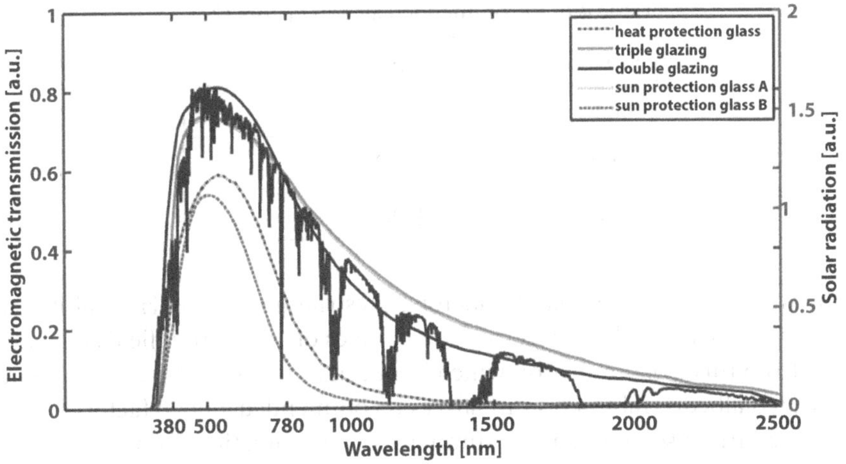

Figure 4.16 Spectral Transmission of common isolation and heat protection window glasses (source: Uniglas, 2010) and spectral solar irradiance AM 1.5 (axis right, qualitative).

Reflection

One of the most important reasons to use simulation programs for the evaluation of photovoltaic converters within buildings is the high contribution of indirect irradiance to the available irradiance. As can be seen from this section and during simulation, this is especially true for the meaningful modeling of the reflection coefficient. This requires determining the expected number of reflections, as a ray passes until it is absorbed completely and converted to heat radiation. The European Norm EN12464-1 suggests the values from Table 4.5 as typical for the degree of reflection for elements in buildings.

4.4.4 Indirect Contributions of Solar Radiation

Solar radiation interacts with its environment, thus reflected, transmitted, and absorbed. This is one of the reasons why solar irradiance can exceed

Table 4.5 Reflection coefficient following EN12464.

Element	ρ_r
Ceiling	0.6-0.9
Side wall	0.3-0.8
Desktop	0.2-0.6
Floor	0.1-0.5

Table 4.6 Recommended albedo factors [22].

Surface	ρ_r
Snow	0.7..0.9
Sand	0.25..0.35
Forest, field	0.1..0.2

the well-known theoretical limits for terrestrial irradiance. High altitude, desert or maritime locations can be examples of application fields benefiting from this. Solar radiation which has been reflected in the environment is called albedo. Table 4.6 summarizes recommended albedo factors [22].

The circumsolar radiation summarizes the radiation from the radial surroundings of the sun. Most of this radiation results from interaction with aerosols within the atmosphere. Thus, the circumsolar radiation varies with season, geography, and meteorological and local factors, such as industrial smog. A well-established model for circumsolar radiation is the semi-empirical Perez model [23]. It covers four different scenarios for the reflection of the sky ranging from clear to cloudy conditions.

4.4.5 Final Results and Limits of Analytical Models

The analytical approach discussed in this section can be helpful to obtain first approximations for general considerations. With a spectrally constant transmission coefficient of 0.6 and a maximum irradiance through direct solar radiation between 169–932 Wm^{-2}, a maximum irradiance between 210–442 Wm^{-2} can be expected directly behind the windows of the modeled rooms. The modeling of concrete and varied scenarios requires solving numerous equations, rendering the use of a simulation program necessary (see [24]). The following sections introduce the principles of the simulation programs used in this chapters. These programs are classical in their approach and based on the physical relations and models discussed above.

4.5 Simulations

4.5.1 Ray Tracing: Fundamental Principles

There are two concepts for the simulation of irradiance conditions: ray tracing and radiosity. In ray tracing, the paths of individual rays are modeled (traced). Ray tracing is the tracking of single rays after emission. Radiosity first models the irradiance of the whole setting and then adds the effects of an illumination to it. The general equation of radiosity models the radiosity $B(x)$ on infinitesimal partial surfaces dA_i (Eq. 4.26). $B(x)$ on a point x consists of intrinsic radiation E, the reflection factor ρ_r, the distance r and the angle Φ between the partial surfaces x and x'. The relation $V(x,x')$ in Equation 4.28 indicates whether a ray gets from location x to x'. All surfaces are assumed to be complete diffuse Lambertian reflectors, thus spectral characteristics of materials and mirroring reflections are neglected.

S denotes all surfaces taken into consideration and the partial areas dA_i are of infinitesimal size. This leads to an infinite number of equations. Therefore, radiosity models are generally combined with the finite element method (4.28). In addition, it is assumed that all partial areas A_i are planar and that B_i and ρ_i are constant on these areas. This yields a simplified equation with a form factor

F_{ij}. This factor assesses how much of the radiation energy received by is remitted and thus available for lighting applications.

$$B(x) = E(x) + \rho(x) \int_S B(x') \frac{1}{\pi r^2} \cos\phi_x \cos\phi'_x \; V(x,x') dA' \quad (4.25)$$

$$B_i = E_i + \rho_i \sum_{j=1}^{n} B_j F_i j, \quad (4.26)$$

$$1 \leq i \geq n, \quad (4.27)$$

with

$$V(x,x') \begin{cases} 1 & \text{for x visible from x}', \text{otherwise} \\ 0 & \end{cases}$$

The alternative simulation method, ray tracing, provides more detailed models of reflection. Ray tracing programs track individual rays, until they have reached a predefined number of reflections on each object or if their power falls below a certain power limit. Each material has a characteristic spectrum of absorption. Materials within ray tracing programs, such as Radiance, can be assigned with reflection properties. The foundation of ray tracing programs is the conservation of energy. For this reason, they are used for the physically correct simulation of radiation [14].

4.5.2 Radiance

This section introduces the ray tracing program Radiance [14]. Its theory and a practical application for the investigated rooms are presented. Although simulation programs differ in details, much of the theory and practical implications are valid for most of the ray tracing software available.

The received radiation on an investigated surface consists of any intrinsic radiation on the surface (such as for the sun), the received direct radiation from radiation sources in the environment, for example from installed illumination, and the reflection behavior of the surrounding. This is described in the following equation, which is the base of the calculation within Radiance:

$$L_r(\theta_r,\phi_r) = L_e + \int\int L_i(\theta_i,\phi_i) f_r(\theta_i,\phi_i,\theta_r,\phi_r) \left|cos\theta_i\right| \sin\theta_i d\theta_i d\phi_i, \quad (4.28)$$

where

L_i = incident radiation,
f_r = bidirectional reflectance-transmittance distribution function BRTDF,
L_e = intrinsic radiation of the object [25],
L_r = received radiation

The finally calculated resulting radiation L_r consists of the self-emitted radiation L_e of the object and the integral of the function of the illumination distribution of the incoming radiation L_i and the function of the surface behavior f_r. Here, the fraction of direct radiation L_i is influenced by the local distribution and intensity of the radiation sources, and the fractions of indirect radiation are a function of the properties of the surroundings f_r. The function f_r is called bidirectional because it models incident and re-emitted radiation.

The BRDTF f_r describes the radiation scattering properties of a surface. Thus, the definition of materials in Radiance directly influences the BRDTF. For an optically isotropic material, the BRTDF is

$$f_r(\theta_i,\phi_r,\theta_r,\phi_r) = \frac{\rho_d}{\pi} + \frac{\rho_s}{\sqrt{(cos\theta_i\theta_r)}} \frac{e^{-\tan^2\frac{\delta f}{\sigma^2}}}{4\pi\sigma^2}, \quad (4.29)$$

with the standard distribution σ, the incident angle δ_f, the diffuse reflection p_d and the mirroring reflection ρ_s. The detailed calculations can be found in Larson *et al.* [14].

4.5.3 DAYSIM

DAYSIM is a ray tracing program for daylight simulations based on Radiance. Figure 4.17 shows the basic structure of such a simulation. DAYSIM models the dynamics of solar radiation by combining measurement data from real weather stations with sky models. In general, measurement data are available as hour values. For finer resolution up to minutes, DAYSIM generates data based on the stochastic, autocorrelated models from Perez and Skarveith and Olsen [23].

The Perez Sky model generates sky conditions based on weather data (see Section 4.4.4, and [23]) To that aim, two submodels are combined: The Perez model of efficiency of illumination and the model of the distribution of illumination at the sky.

The input parameters of the Perez model of efficiency of illumination are the zenith angle, the altitude above sea level, the direct and diffuse irradiation of the sun and the amount of precipitation water in the atmosphere.

The model of illuminance distribution is based on the albedo, the level of brightness of the horizon, the gradient thereof, the relative irradiance and the diffusion of the circumsolar region. The Perez model, which is also

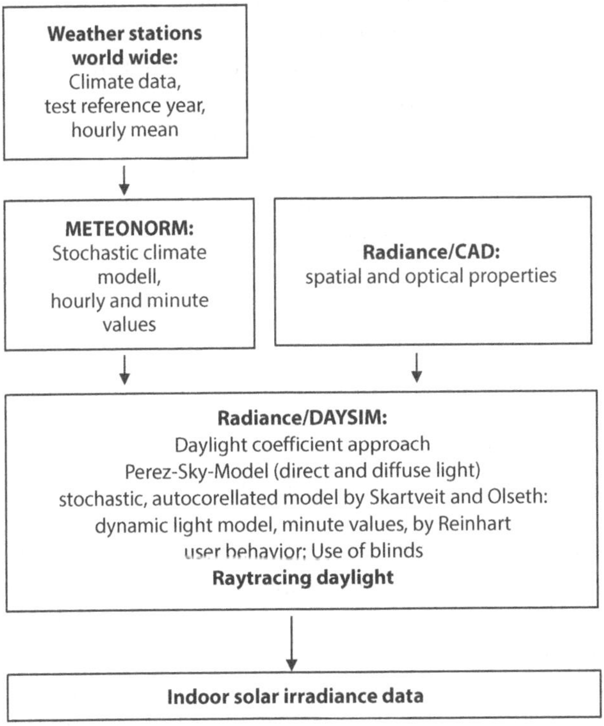

Figure 4.17 Structure of a DAYSIM simulation.

implemented in Radiance, does not account for short-term fluctuations. Thus, Skarveith and Olsen developed the model named after them, which factors this aspect in. Based on this, Reinhart and Walkenhorst extended the model of short-term dynamics further to a resolution of minutes [15]. The extended model is integrated in DAYSIM. In addition to that, DAYSIM features user models for the use of electric illumination and shading as well as the calculation of the daylight coefficient (see below).

4.5.4 Calculation Methods and Parameters

This section introduces the main calculation methods for direct and indirect radiation in Radiance and DAYSIM. It explains the calculation of the daylight coefficient. The simulation parameters used as the exemplary simulation in this chapter are presented.

Calculation of Direct Radiation

Direct radiation L_i is mostly influenced by the position, type, and geometry of the radiation sources. Further calculations of direct radiation in Radiance contain, depending on the simulation model, so-called virtual light sources resulting from reflections and weighting of light sources when using multiple light sources (selective shadow testing).

For the calculation of incident indirect radiation L_i, a planar, homogeneous radiation source is assumed. Large or geometrically complex radiation sources are divided into smaller sources in order to avoid errors due to incorrect solid angles or partial shading. This approach is controlled by the parameter *ds*.

Direct Subsampling "ds"

"*ds*" subdivides radiation sources for additional calculations. This is, for example, necessary for significantly bent surfaces or very extended radiation sources. Sunlight has an incident angle of 0.5 in DAYSIM. In order to lower the required calculation time, "*ds*" is set to 0 in DAYSIM. A value of 0 means that a radiation source will not be subdivided.

The related parameter direct jitter "dj" subdivides radiation sources for additional calculations. This is, for example, necessary for significantly bent or very extended radiation sources. The incident angle of sunlight in DAYSIM is 0.5°. Thus, direct jitter "dj" is set to 0 in DAYSIM, in order to lower calculation times. With a value of 0, a radiation source is not allowed to be subdivided.

Finally, the parameter direct jitter "dj" controls, how many rays per subarea in accidentally distributed x-y-directions are calculated.

Direct Threshold "dt"

In order to optimize processing times, the parameter direct threshold "dt" evaluates radiation sources regarding their visibility and radiant intensity. If the radiation source undercuts a certain value, its influence on the investigated surfaces is considered to be negligible and the radiation source is dismissed from the calculations. The maximum allowed error for this is defined by the parameter direct certainty "dc". In DAYSIM, "dt" is set to 0 automatically for the calculation of the daylight coefficient *DC*, as only effects of sunlight are investigated. The treatment of virtual radiation sources is defined with the parameters direct relay "dr" and direct pretest "dp."

Calculation of Indirect Radiation

Radiance uses two approaches in order to speed up the calculation of indirect radiation: "Specular Sampling" and "Indirect Irradiance Caching." Specular Sampling covers all diffuse radiation, that is non-Lambertian radiation, such as refracted or reflected light. The calculation effort for non-ideal mirroring reflection is minimized by using a so-called selective Monte Carlo importance sampling [26].

The Indirect Irradiance Caching increases the performance by finding the mean indirect irradiation for a point. Between investigated points, the indirect irradiation is interpolated based on the results. To that aim, a large number of rays are emitted in random directions within the hemisphere of the investigated points. The number of rays is defined by the parameter "ad" (ambient divisions). As this is a statistical approach, the accuracy of the calculation of indirect irradiation increases with this number. The direction of the emitted rays is defined by the following Monte Carlo inversion procedure:

$$E = \frac{\pi}{MN} \sum_{j=0}^{M-1} \sum_{k=0}^{N-1} L_{j,k} \tag{4.30}$$

where

$L_{j,k}$ denotes the indirect irradiation in the direction $\vartheta_j \Phi_k = \left(\arcsin\sqrt{\frac{j+X_j}{M}}, 2\pi\frac{k+Y_k}{N} \right)$ equally distributed random variable in the range from 0 to 1, and $M.N$ the total number of samples is $N \approx \pi M$.

Ambient Bounces "ab"

Ambient bounces "ab" define how often the reflection of a ray is traced and, by this, how many indirect equations have to solved. Thus, this parameter is crucial for the required processing time. For a room with simple reflection conditions, a value of 5 is considered to be sufficient.

Ambient Accuracy "aa"

The parameter "aa" defines the maximum allowed error in the calculation of the interpolation of the indirect irradiation. Together with ambient resolution "ar," it defines the maximum spatial resolution S_{min} of the interpolation of the indirect irradiation (4.32):

$$S_{min} = \frac{D_{max} aa}{ar} \tag{4.31}$$

D_{max} represents the largest value of the dimensions of the investigated scenario, "ar" specifies the minimum distance of a point and is analog to the finest grid resolution of a Radiosity program. Scenarios thus should be kept as small as possible. Furthermore, the processing time increases with decrease of *ar*. In the exemplary simulation of the north office, the maximum dimension D_{max} is 10 m, so with $aa = 0.1$ and $ar = 256$ the maximum spatial resolution becomes 3.9 mm. The results presented in the following sections have been obtained using these settings.

4.5.5 Daylight Coefficient in *DAYSIM*

In general, a ray tracing program performs a unique ray tracing for each sky condition throughout the year. In order to reduce the required processing time, DAYSIM enables the simulation of a whole year by using the daylight coefficient $DC_\alpha(x)$ approach. This approach calculates the daylight coefficient for each investigated point following (4.33). The daylight coefficient DC identifies possible position of the sun during the year. To that aim, the sky is divided into segments of spherical angles S. The daylight coefficient DC for a position x in a space then follows Reinhart [27]:

$$DC(x) = \frac{E(x)}{L \ S}, \tag{4.32}$$

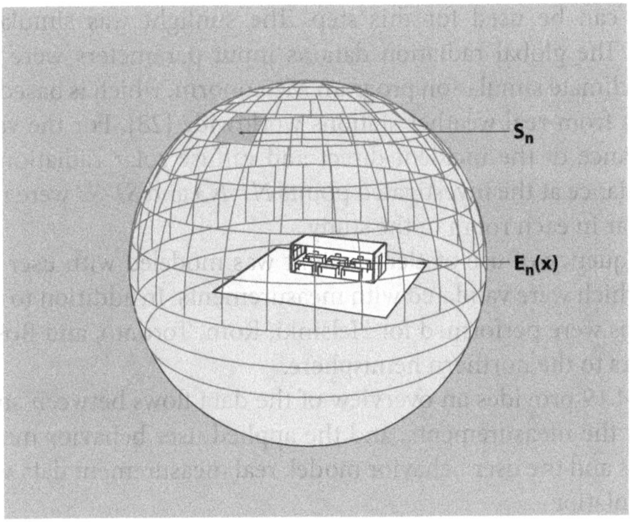

Figure 4.18 Daylight coefficient approach in DAYSIM (see Eq. 4.33). (Copyright: [27]).

with the illuminance E and the luminance of the sky L. Figure 4.18 depicts the approach. In the radiation simulation, the determined daylight coefficient DC then is combined with the sky model of the simulation interval of seconds or minutes.

4.5.6 Environmental Parameters

Indoor radiation consists of four components, which are subdivided into the following variables:

- characteristics of the room: geometry, reflection coefficient, transmission coefficient
- artificial light: position, emission characteristics, spectral irradiance, lamp ballast factor
- solar radiation: geographical position, temporal position within the year, temporal position within the day (incident angle), meteorological effects (shading), albedo
- user: presence, frequency and endurance of usage of artificial lighting and sun shields.

In this section, the geometrical and optical properties of a room and the artificial light have been modeled in Radiance. Alternatively, architectural programs can be used for this step. The sunlight was simulated with DAYSIM. The global radiation data as input parameters were obtained using the climate simulation program Meteonorm, which is based on measured data from real weather stations worldwide [28]. For the validation, the irradiance of the incident direct and diffuse solar radiation and the total irradiance at the investigated points *N1-N3* and *S1-S3* were measured for one year in each room of the study.

The frequency of use of electric light was modeled with user behavior models, which were validated with measurements. In addition to Freiburg, simulations were performed for Helsinki, Rom, Toronto, and Boston. The study refers to the northern hemisphere.

Figure 4.19 provides an overview of the data flows between simulation programs, the measurements, and the applied user behavior models. For validations and the user behavior model, real measurement data were used in the simulation.

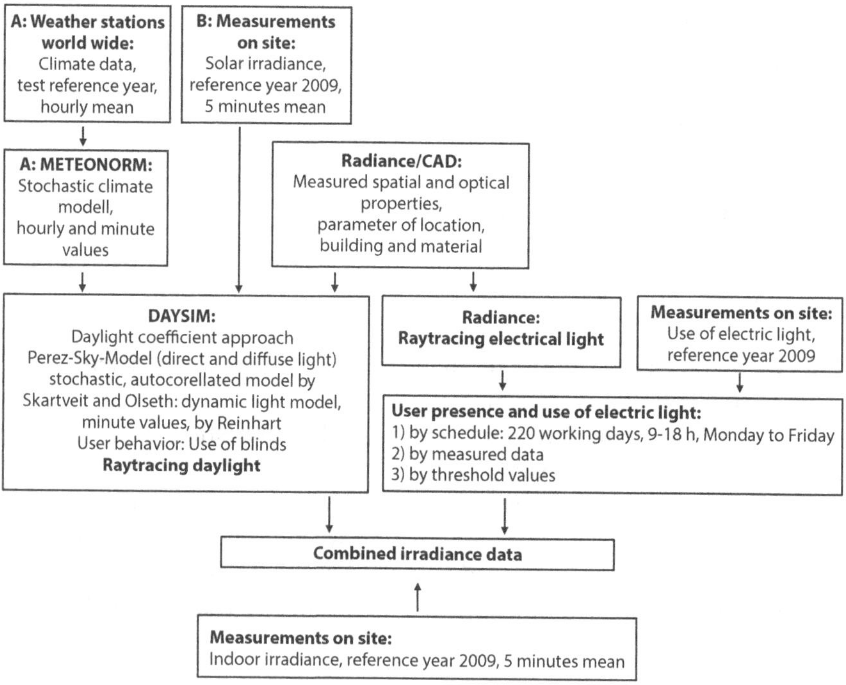

Figure 4.19 Data flow of the simulation of the combined irradiation from solar and artificial sources and for validations.

4.5.7 Model Parameters

Room

The north office has 70 m^2 with a ceiling height of 3.87 m. The south office has 38 m^2 with a height of 3.28 m. The position and size of all furniture was measured and applied in the room models. Annex.2 provides some detailed example of the room model of the north office in Radiance.

The reflection parameters of the most important surfaces in the room were measured following the procedures described in [14] and implemented in the simulation. Table 4.7 compares the measured values to the recommended values for illumination planning in the European Standard EN12464-1 [4]. The reflection coefficients of the surfaces of the rooms are medium to high for the walls, ceilings and work surfaces, and low for the floors when compared to the European Standard.

Artificial Lighting

In both offices, fluorescent tubes of daylight spectrum of the type OSRAM Lu-milux 58W/840, cool white (5200 Lumen) were installed. The emission characteristics and the efficiency factor of the luminaire can often be found in the database of the Illuminating Engineering Society of North America (IES). In this example, the original *ies*-files were not available, and data of similar lamp types were used for the simulations in Radiance. The efficiency factor of the luminaires in both offices was set to 0.64.

Table 4.7 Comparison of the applied reflection coefficients to the recommended values of the European Standard EN12464 for illumination planning.

Element	EN 12464	North office*	South office*
Ceiling	0.6-0.9	0.70	0.70
Wall	0.3-0.8	0.85	0.84
Working surfaces	0.2-0.6	0.47	0.47
Floor	0.1-0.5	0.24	0.23

*Arithmetic mean, measured following [14], n = 25. The coefficient of the ceiling was estimated based on experience.

Daylight

The DAYSIM simulation in this example used the room models described above, climate data from Meteonorm, the transmission coefficient T_n of the windows, and the albedo values listed above.

The standard procedure in DAYSIM is to use the manufacturer data of the investigated windows or measured data of the real transmission in the visible range. Based on the measurement results the value required for the simulation is calculated [14].

In contrast, IPV applications require the complete wavelength interval of the solar radiation. Furthermore, the windows investigated in the example were produced in 1996. The original transmission data were unavailable, and the windows did not meet the current technological standards. For these reasons, two approaches were pursued in order to model the window transmission in the simulation.

The simulation models and results discussed in the following sections have been obtained with an assumed g-value of 0.6. This is equivalent to a transmission coefficient of 300–2500 nm and covers most of today's common types of glazing. As mentioned in Section 4.4.3, sun protection windows need to be considered separately.

Subtracting 10% for dirt and another 15% for shading from window frames finally yields a transmission coefficient T_n of 0.459 with a transmissivity coefficient $t_n = 0.5426$. The transmissivity coefficient t_n was calculated with a relative refractive index of 1.52 for glass and the reduced formula from [14] (see Annex.3). For the simulation with data from measured solar incident radiation at the end of this chapter, a semi-empirical approach was used.

In accordance with the literature, the albedo was assumed to be 0.2 [22, 27].

User Behavior

Three approaches were investigated for the modeling of the frequency and the duration of use of artificial lighting.

1. Simple time of use model: Assumed presence of users from Monday until Friday from 9 h until 18 h, 220 working days, in rough accordance with the user habits of the investigated offices. The presence of users is assumed to imply the ue of electric light.
2. Empirical: The indoor irradiation was measured permanently, so the use of electric light was detected. For every measured artificial light in the reference year, simulated artificial light was added in the simulation.

3. Threshold value model: Definition of threshold based on the measured radiometric values in the reference year. Artificial light is switched on when a certain irradiation is undercut. The irradiation value is determined from the measurements. In this approach, the spectral sensitivity of the human eye adds an uncertainty to the results.

The "simple dynamic shading" model from DAYSIM was used for the user behavior model of the utilization of sun protection devices (shading). This model assumes from a certain level of glare from sunlight the existence and use of a simple sun protection with a transmission of 25% [27].

4.5.8 Results

Required Complexity of a Simulation Model

In real applications, rooms do not only consist of basic furniture. They also contain various details, such as plants, shelf space, or smaller pieces of furniture, all which rendering a simulation model very elaborate. In addition to that, the statistical portion of the simulation increases due to the rising portion of indirect irradiance. For future applications, the minimum required level of complexity and the resulting minimum effort for a simulation model are of interest, with sufficient accuracy for the design of an IPV system.

These questions were studied for the three sensors in the north office. Figure 4.6 shows a photo of the office: a room with many complex forms and reflecting elements. Table 4.8 contains the simulation results for the unfurnished room (Radiance model "geometry", Figure 4.20), for a simplified room "basic furniture" with desks and cupboards (Figure 4.21) and for a model "details" with small furniture, plants, and shelf spaces (Figure 4.22).

Table 4.8 Influence of the complexity of the simulation model on the result for the irradiance.

Model	Irradiation N1 in W/m²	Irradiation N2 in W/m²	Irradiation N3 in W/m²
Geometry	1.44	2.55	0.69
Basic furniture	1.29	2.04	0.49
Details	1.19	2.00	0.43

Figure 4.20 The north office in Radiance: Model "Geometry," hemispherical view.

Figure 4.21 North office in Radiance: Model "Basic furniture," hemispherical view.

Figure 4.22 North office in Radiance: Model "Details," hemispherical view.

DAYSIM uses the same ray tracing as Radiance. Therefore, the validation of the Radiance model is valid for DAYSIM, too, besides the influence of weather conditions. In order to obtain a constant radiation source, that is independent from the daylight, only the electric light was simulated and the ray tracing was performed in Radiance. The applied simulation parameters are listed in Table 4.9. The sensor *N2* is placed on a roll-fronted cabinet directly under the luminaires. *N1* is oriented toward the window at a 10 m distance. The sensor *N3* faces the opposite wall and receives no direct radiation. Table 4.8 summarizes the results.

As expected, the simulation results for an empty room differ from the results obtained for a furnished room. The difference between the results for the simple model from Figure 4.21 and the complex model from Figure 4.22 varies between 2% for *N2*, which receives mostly direct radiation, and 14% for *N3* with mostly indirect radiation. In DAYSIM, only rounded numbers are implied for model assumptions and the results between both

Table 4.9 Values of the applied simulation parameters.

ab	ad	as	ar	aa	ds
7	1024	128	256	0.1	0.1

models were identical. Depending on the application, the simple model thus can be sufficient for the dimensioning of IPV.

Characteristic Values of Solar Radiation in Buildings

In mass production items, a product powered with photovoltaics needs to be reliable worldwide. This section investigates the effect of orientation towards the window, and the effect of the latitude and of the orientation of the window itself on the annual mean of the irradiance. To that aim, three rows of virtual sensors with a height of 1 *m* were simulated in the unfurnished model of the north office without a glass dividing wall in DAYSIM. Each orientation vector was simulated separately. The x-row is orientated toward the wall, the y-row toward the window and the z-row toward the ceiling. The distance to the window started with 0.5 m and increased in 1 m increments.

The Figures 4.24 and 4.26 depict the orientation of the measurement points in the north office, and in the south office, respectively. For the sensors in the south office, a daily mean maximum irradiance of 56 Wm^{-2} (SI) and 108 Wm^{-2} (S2) was measured, for the north office between 1.5 Wm^{-2} (*N2*) and 9.5 Wm^{-2} (*N3*). The annual mean was 0.87 Wm^{-2}, 0.69 Wm^{-2}, and 3.03 Wm^{-2} for sensors in the north office. In the south office, the annual mean for the individual sensors was 16.74 Wm^{-2}, 26.1 Wm^{-2}, and 43.45 Wm^{-2}. The maximum instantaneous values were 444 Wm^{-2}, 528 Wm^{-2}, and 471 Wm^{-2}. In the north office, the maximum instantaneous values were 6 Wm^{-2}, 9 Wm^{-2}, and 120 Wm^{-2}. Table 4.10 summarizes the results.

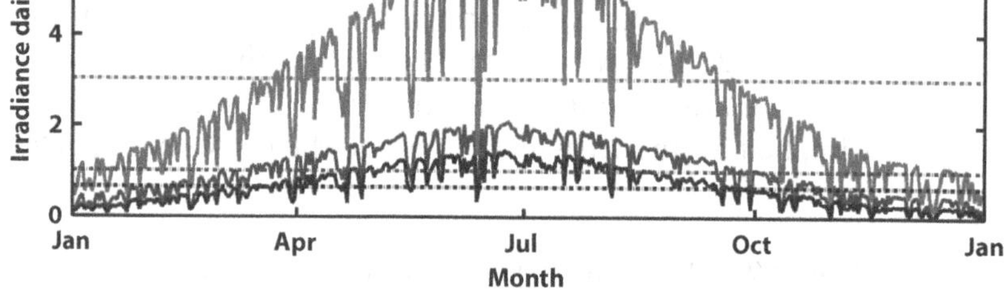

Figure 4.23 Mean solar irradiance for sensors in the north office, detailed, DAYSIM model, location Freiburg. $\sigma_{N1} = 0.57$, $\sigma_{N2} = 0.40$, $\sigma_{N3} = 2.05$, mean and maximum values when compared to the south office (see Table 4.10).

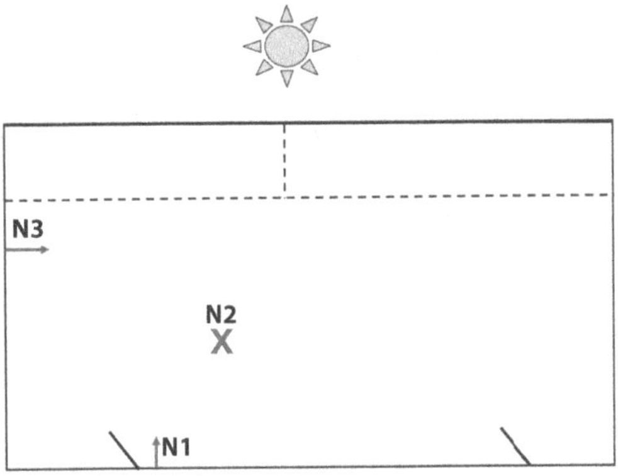

Figure 4.24 Schematic of the orientation of the measurement points in the room, north office.

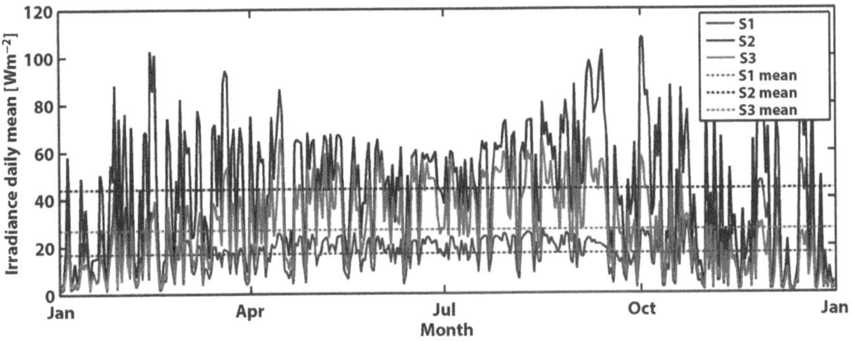

Figure 4.25 Mean solar irradiance for sensors in the south office, DAYSIM model, location Freiburg. $\sigma_{S1} = 10.39$, $\sigma_{S2} = 26.32$ $\sigma_{S3} = 17.60$. Mean and maximum in comparison to the north office (see Table 4.10).

Figure 4.27 shows the results for the north office from Figure 4.23 in the same scale as required for the south office in Figure 4.25. In the model, the window was either orientated toward the north, as in Figure 4.28, or toward the south, as in Figure 4.29. In order to investigate the influence of the latitude, Rome and Helsinki were chosen as locations with a difference in latitude of 19°. As a result, the influence of the orientation of the window itself exceeds the influence of the latitude. The simulation results for Helsinki remained under the results for Rome with a factor of about 0.75. For both latitudes, the irradiance in a room with a north window only reached 25% of the irradiance received in a room with a south window. For all simulations, the irradiance of the sensors facing a wall remained on a base level.

Table 4.10 Mean and maximum values of simulated solar irradiance for Freiburg in the north and south office.

Sensor	$\bar{E}_{eannual}$ Wm^{-2}	$\hat{E}_{eannual}$ Wm^{-2}
N1	0.87	6
N2	0.69	9
N3	3.03	120
S1	16.74	444
S2	26.10	528
S3	43.45	471

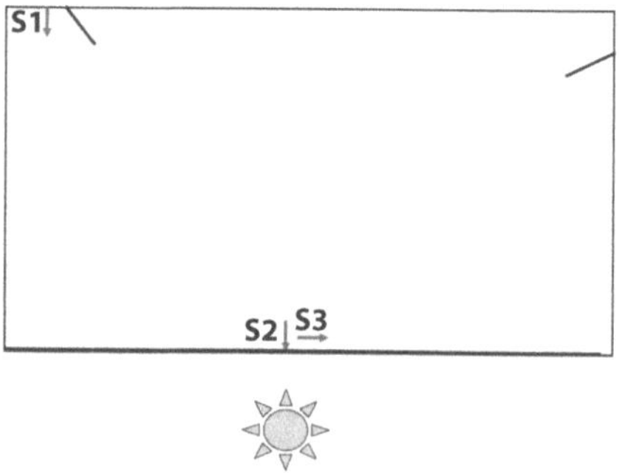

Figure 4.26 Schematic of the orientation of the measurement points in the room, south office.

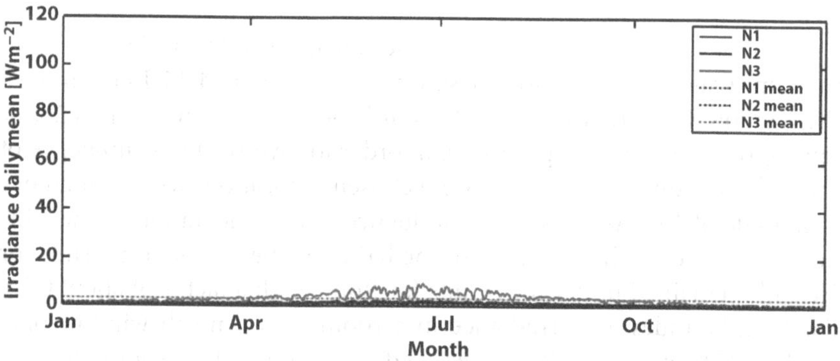

Figure 4.27 Mean solar irradiance for sensors in the north office, scaling in comparison to south office, DAYSIM model, location Freiburg. $\sigma_{N1} = 0.57$, $\sigma_{N2} = 0.40$, $\sigma_{N3} = 2.05$.

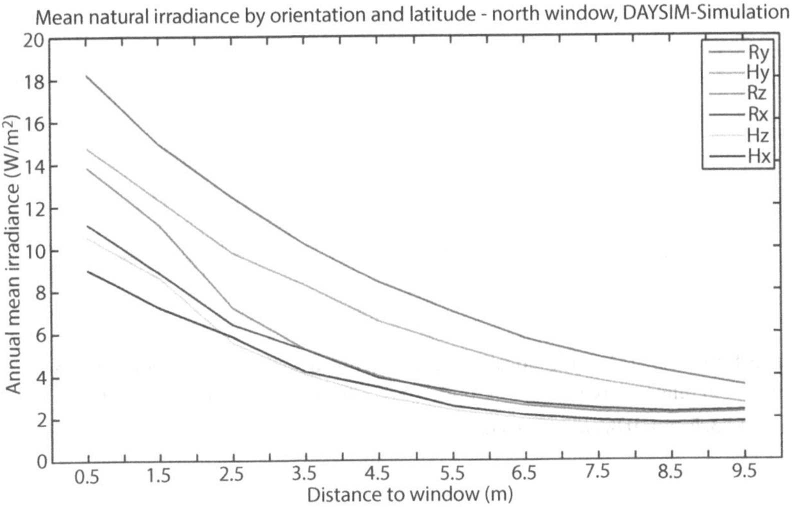

Figure 4.28 DAYSIM simulation with a mean solar irradiance in Wm^{-2} for an empty room with a north window. 'R' stands for Rome, 'H' for Helsinki, and 'x', 'y', and 'z' are the orientation vectors.

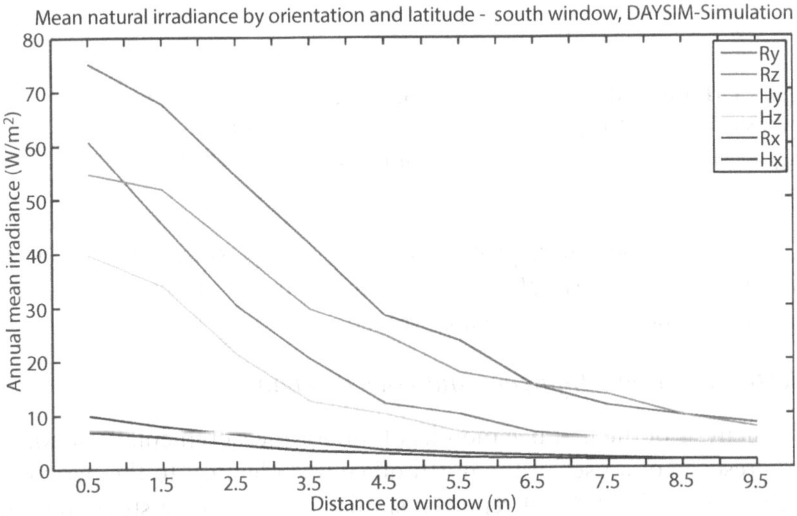

Figure 4.29 DAYSIM simulation with a mean solar irradiance in Wm^{-2} for an empty room with a south window. 'R' stands for Rome, 'H' for Helsinki, and 'x', 'y', and 'z' are the orientation vectors.

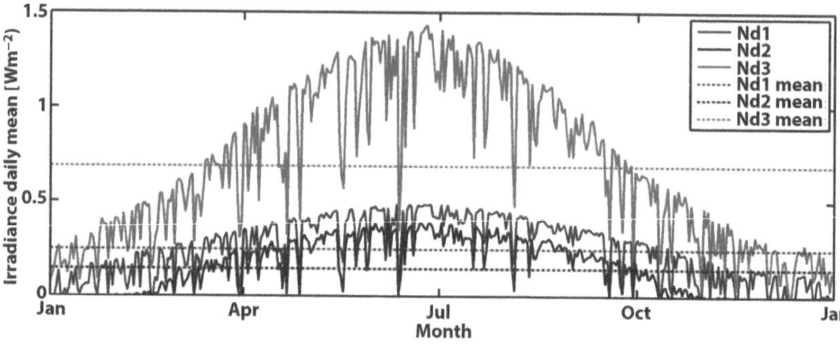

Figure 4.30 DAYSIM model, location Freiburg, mean solar irradiance for sensors in the north office, glare protection model, $\sigma_{Nd1} = 0.15$, $\sigma_{Nd2} = 0.14$, $\sigma_{Nd3} = 0.41$, $\hat{E}_{Nd1} = 1.73, \hat{E}_{Nd2} = 0.87, \hat{E}_{Nd3} = 4.34$.

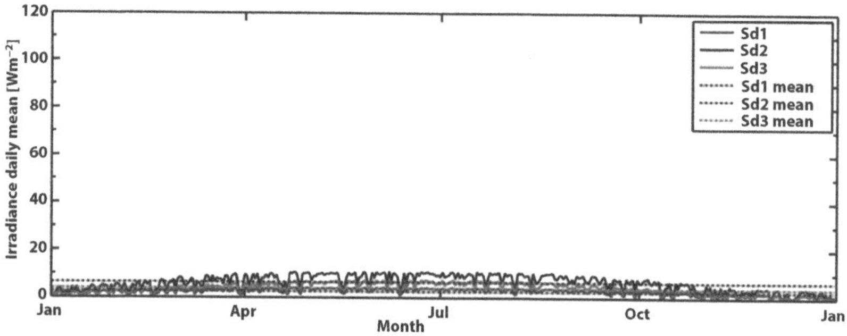

Figure 4.31 DAYSIM model, location Freiburg, mean solar irradiance for sensors in the south office, glare protection model, $\sigma_{Sd1} = 117$, $\sigma_{Sd2} = 3.17$, $\sigma_{Sd3} = 2.14$, $\mu_{Sd1} = 2.54, \mu_{Sd2} = 6.35, \mu_{Sd3} = 3.87, \hat{E}_{Sd1} = 4.30, \hat{E}_{Sd2} = 11.14, \hat{E}_{Sd3} = 7.55$.

In some regions, glare protection might change simulation results significantly. The Figures 4.30 to 4.35 summarize the results for the two offices with and without the use of glare protection.

Electric Lighting—Influence and User Behavior

The analysis of the first five months of indoor measurements showed, that the presence of a user in the north office was identical to the permanent use of electric light. Hunt reported similar results from a study in school buildings [29]. The real working times exceeded the ones obtained from the simple scheme model which assumed working hours from Monday until Friday from 9–18 h.

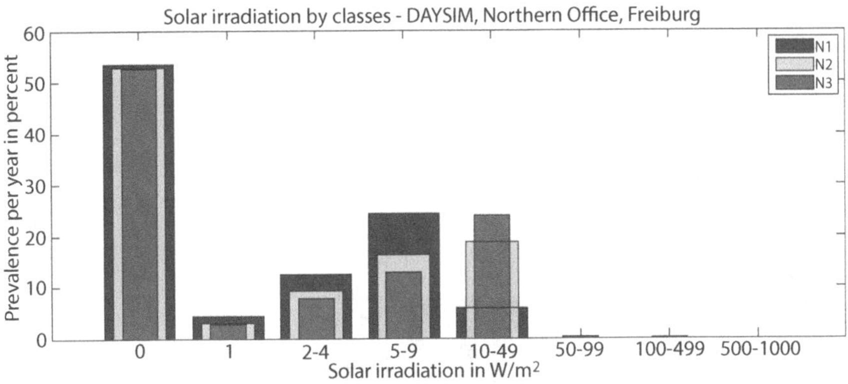

Figure 4.32 Histogram of solar irradiance: DAYSIM model, north office, Freiburg.

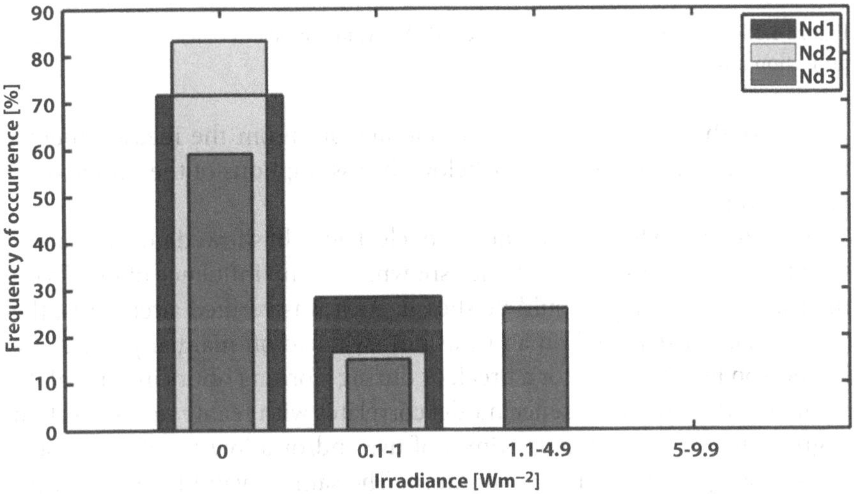

Figure 4.33 Histogram of solar irradiance - *DAYSIM*-model, north office, Freiburg, glare protection model.

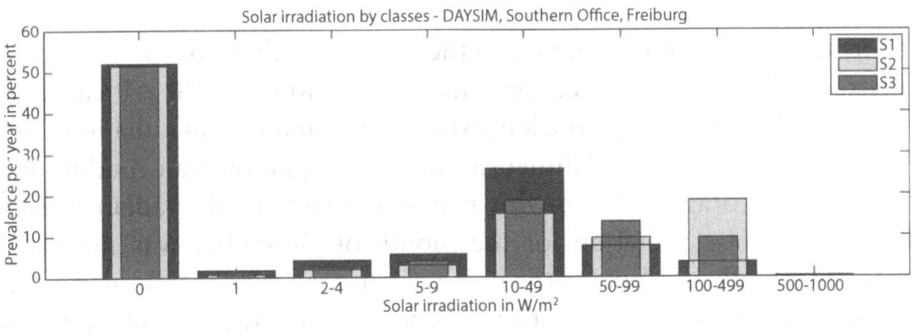

Figure 4.34 Histogram of solar irradiance: DAYSIM model, south office, Freiburg.

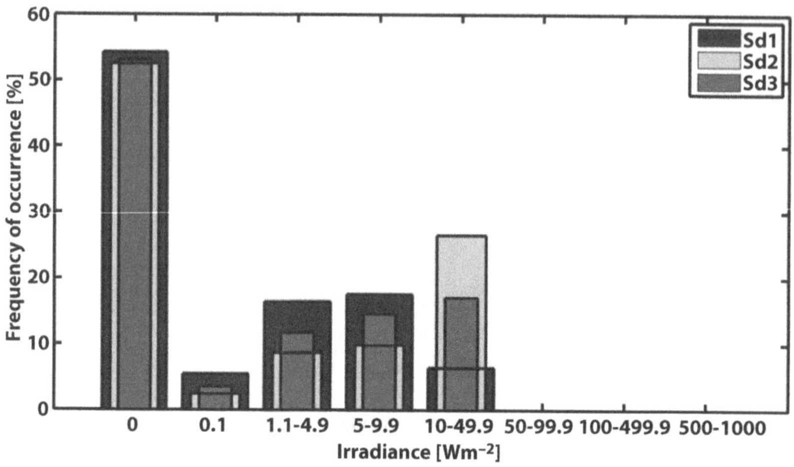

Figure 4.35 Histogram of solar irradiance: DAYSIM model, south office, Freiburg, glare protection model.

The south office was used from one person. From the measurements, the use of electric light was 50% below the assumptions of the simple time of use model.

It is interesting to see, that the use of electric light showed no correlation with the measured solar irradiance shown. Also, no influence of breaks on the use of electric light could be shown. As it was verified afterwards, the light remained switched on and was not switched off manually, when the last person left the room for a break or during working hours in general. In summary, the simple presence model correlates with reality up to a certain degree. Offices with irregular times of use and/or a low number of users need an adaptation to these conditions. The same is valid for rooms with automatic lighting and potentially frequent episodes without electric light.

Time of Autarky with December as Example

The irradiance for an orientation to the south and close toward a window can exceed 500 Wm^{-2} for short periods of time. However, the critical value for system dimensioning is the longest period with minimal radiation. This period defines the required dimensioning for the photovoltaic module and the required storage unit. For this reason the combined irradiance from electric and solar radiation for the month of December was simulated for the north office. Also, for the user presence, a worst case scenario was assumed with an assumption of a weekend followed by two bank holidays. Figure 4.36 shows the frequency of different classes of irradiance from

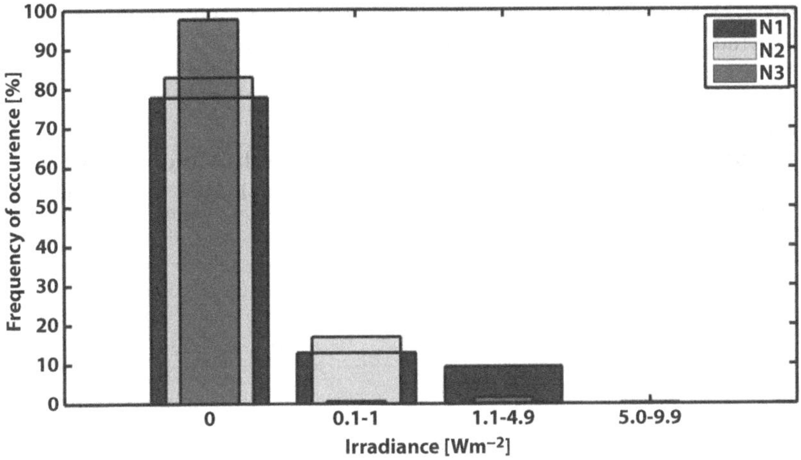

Figure 4.36 Histogram of solar irradiance: DAYSIM model, north office, Freiburg, December.

daylight. For Figure 4.37 electric light following the simple presence model was added. For sensors with orientation toward the ceiling, the contribution of electric light was 50%. This is a very important conclusion for all applications on window desks.

Table 4.11 Summarizes the relation between orientation of the receiving area and the distance to the window for both an empty and a furnished room. As a result, the influence of the orientation vector exceeds the influence of the location vector. In addition to that, the influence of the

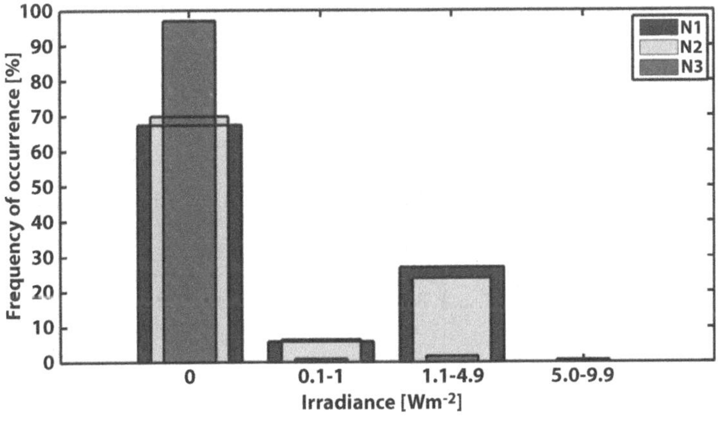

Figure 4.37 Histogram of combined irradiance: DAYSIM + Radiance + user-model, north office, Freiburg.

Table 4.11 Influence of location and orientation vector on the annual mean solar irradiance $\overline{E}_e$ for two rooms in Helsinki and Rome.

Distance to window in m	**North-Rome $\overline{E}_e$ in Wm^{-2}**	**North-Helsinki $\overline{E}_e$ in Wm^{-2}**	**South-Rome $\overline{E}_e$ in Wm^{-2}**	**South-Helsinki $\overline{E}_e$ in Wm^{-2}**
Orientation vector x=l, y=0, z=0				
0.5	9.69	7.81	8.63	6.03
1	7.67	6.25	6.88	5.20
1.5	5.57	5.07	5.59	3.88
2	4.54	3.65	4.16	2.89
2.5	3.38	3.00	2.97	2.34
3	2.84	2.22	2.45	1.73
3.5	2.36	1.85	2.06	1.53
4	2.12	1.62	1.65	1.32
4.5	1.97	1.50	1.32	1.27
5	2.03	1.55	1.29	1.21
Orientation vector x=0, y=l, z=0				
0.5	15.83	12.1	11.11	47.45
1	12.95	10.67	9.25	44.96
1.5	10.79	8.52	7.39	35.35
2	8.85	7.17	6.22	25.69
2.5	7.28	5.69	4.93	21.39
3	6.07	4.70	4.08	15.38
3.5	4.11	3.84	3.34	13.35
4	4.22	3.25	2.83	11.75
4.5	3.59	2.76	2.40	8.44
5	3.05	2.33	2.02	6.47

(Continued)

Table 4.11 Influence of location and orientation vector on the annual mean solar irradiance $\overline{E}_e$ for two rooms in Helsinki and Rome. (*Continued*)

Distance to window in m	North-Rome $\overline{E}_e$ in Wm^{-2}	North-Helsinki $\overline{E}_e$ in Wm^{-2}	South-Rome $\overline{E}_e$ in Wm^{-2}	South-Helsinki $\overline{E}_e$ in Wm^{-2}
Orientation vector x=0, y=0, z=1				
0.5	12.01	9.16	52.70	34.38
1	9.65	6.95	39.44	29.52
1.5	6.26	4.84	26.41	18.53
2	4.56	3.54	17.65	10.81
2.5	3.45	2.63	10.42	8.73
3	2.71	2.06	8.71	5.84
3.5	2.25	1.69	5.76	5.16
4	1.99	1.49	4.65	4.73
4.5	1.89	1.41	4.43	4.02
5	1.97	1.47	4.39	3.85

Height for all vectors 1 *m*, 0,5 m-steps.
Position center of the room, and wall, respectively, for increment in the room.

metereological orientation is at first more important than the geographical position. All of these results are very positive from a system design perspective, as they reduce the required effort for models and enable simple recommendations as rules of thumb for installation places.

Figure 4.38 depicts the receiving characteristic for natural radiation depending on the orientation of the sensor. The simulated location is Freiburg.

4.5.9 Summary and Conclusion

Especially for the simulation of surfaces with an orientation toward a direct radiation source, the modeling of the main objects of the room is sufficient.

The measurement of the use of electric light showed a tendency of the users to permanent use, disregarding the season and the time within the day. Hence, for regular use of a room the simple presence model with times of use is sufficient as user behavior model.

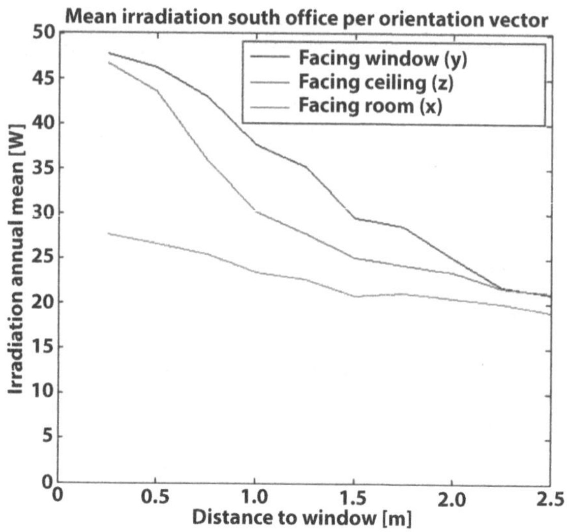

Figure 4.38 Influence of the orientation of the sensor to the radiation source: south office.

4.6 Measurements

4.6.1 Available Measurement Methods

For historic reasons, most of the existing methods to measure indoor optical radiation are photometric and lamps are characterized by their manufacturers. Additionally, most outdoor measurement instruments are too inaccurate for indoor applications due the different indoor spectra and intensities up to four order of magnitudes below outdoor conditions.

Most radiation measurement instruments are based on thermal and photoelectric effects. This sections introduces common instruments.

Pyranometer

Pyranometers (Figure 4.39) are standard instruments for outdoor measurements of solar radiation. These instruments measure diffuse and direct irradiance for a hemispherical field of view based on thermoelectric arrays. Their unamplified output signal is a few μV per Wm^{-2}. The spectral sensitivity is roughly a rectangle function from 310 nm to 2800 nm.

These devices can be equipped with a shadow ring to exclude direct radiation and detect only diffuse radiation.

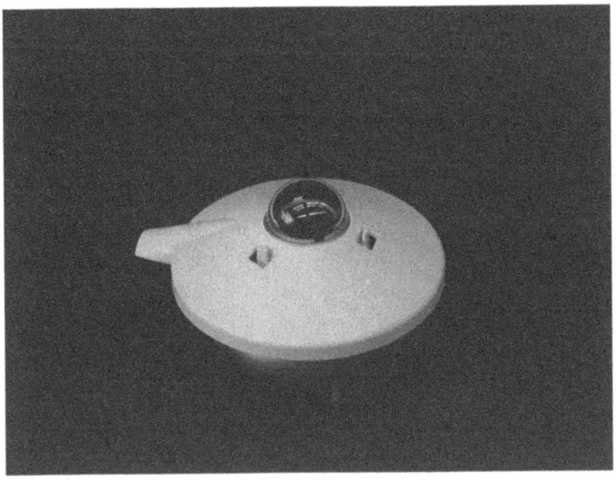

Figure 4.39 Pyranometer to measure the irradiance E_e.

Pyrheliometer

Pyrheliometers use a thermal sensor element. Through a small opening angle and field of view these devices measure direct radiation. In order to measure solar radiation, they need to track it.

Sunshine Duration Transmitters

These instruments measure the endurance of the existence of a certain minimum of solar radiation. In a classical setup, concentrated solar radiation burns a trace into a measurement card. The minimum level to be reached ranges, depending on the device, between 70 Wm^{-2} and 280 Wm^{-2}.

Semiconductor-Based Radiation Sensors

When provided with a sufficient level of irradiance and a sufficient quality of material, the photoinduced short circuit of a silicon solar device is proportional to the received irradiance. Therefore, silicon-based radiation sensors with a cost of about 50–200 Euro are an economic alternative to pyranometers with investment cost of 600–3000 Euro. In addition, the effort for the signal processing is lower due to the higher signal-to-noise ratio. For these reasons, silicon-based radiation sensors are common for the monitoring of photovoltaic and solar thermal outdoor plants. Figure 4.40 shows the exemplary silicon solar radiation sensor. Depending on the customer and the application, the sensors are handheld or installed as

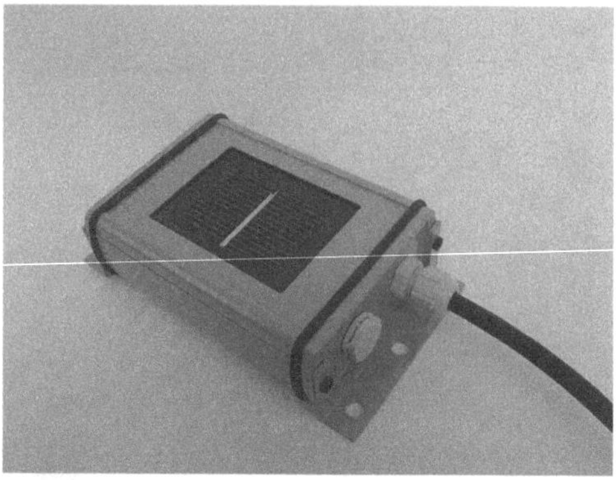

Figure 4.40 Commercial silicon solar radiation sensor for outdoor applications [30].

stationary instruments, and are individually calibrated to electronically compensate for temperature effects.

Measurement made by other semiconductors, such as germanium, follow the same principle and are sensitive in another spectral range.

Luxmeter

Radiation in buildings is optimized for the spectral sensitivity function of the human eye, $V(\lambda)$ (Section 4.2.1). An illumination is measured with luxmeters in Lux. Generally, luxmeters are based on silicon photodiodes, which might be optimized for low intensities. Due to their integrated $V(\lambda)$-filter they are limited to measurements in the visible range from 380–780 nm. As these instruments are nearly the only available standard devices for indoor radiation measurements, they can also be found in indoor photovoltaics. Section 4.2.1 outlines the effect of the spectral mismatch. The consequences for the characterization of indoor photovoltaic devices are explained in Chapter 5.

Spectroradiometer

Spectral irradiance $E_{e\lambda}$ is measured with spectroradiometers, which are based on photoelectric semiconductor arrays and fiber optics. Due to that, they are relevant for calibrations, but mostly unsuited for mobile or long-term measurements. As for all detectors discussed above, the spectral measurement range needs to be adapted to the measurand.

Simulation Programs

As shown in this chapter, various variables, such as geographical position, orientation toward the radiation source, shading, or, for indoor applications, furniture and user behavior, influence the measurable radiation. These variables restrict the usability of the measurement results on the specific measurement situation, which have been obtained with immense effort. For a similar reason, the use of simulation programs, such as PV*SOL or DAYSIM is common practice in the planning of solar plants for outdoor applications. Simulations can replace measurement, depending on the application and the required accuracy.

Possible use and limits of the measurement instruments discussed above are investigated in the following sections with regard to different aspects.

4.6.2 Long-Term Measurements Reference Year

In order to investigate the influence of solar radiation and user behavior, long-term measurements were installed and ran for one year both in the north and the south office. The room and the outside temperatures were recorded. The irradiance E_e was measured at the investigated points *N1-N3* and *SI -S3* with Kipp & Zonen CMP3 pyranometers [31]. At the windows, the incident solar radiation was measured with semiconductor-based detectors of the type depicted in Figure 4.47 [32]. The signal were processed and data were stored in *ADL-MX* dataloggers [33]. The Figures 4.41 to 4.43 depict the installations. The temperature data were measured to investigate the use of thermoelectric generators (see Chapter 2).

As discussed in Section 4.6.4, there are no standard radiometric detectors for indoor applications. Therefore, pyranometers were used for the indoor radiation measurements. With regard to the orientation of the simulation points, most of the pyranometers were installed vertically (Figure 4.44). To the knowledge of the author, this has been the first use and installation of this type.

The measured solar radiation was used as input for the validation simulation. The results of this simulation were compared to the results obtained from the indoor radiation measurements (Section 4.6.3). The Figures 4.45 and 4.46 show the environment of the south office. The north office has a similar situation.

The opposite building causes regular shading and reflection with a strong portion of mirrored radiation. The outdoor connecting hallways are especially complex to simulate. In a classical setup, the solar radiation measurement would need to be performed for an unshaded area, such as on a

Figure 4.41 Measurement installation in the south office with context. A: Pyranometer *S2*, B: room temperature sensor, C: Datalogger, D: Floor temperature sensor.

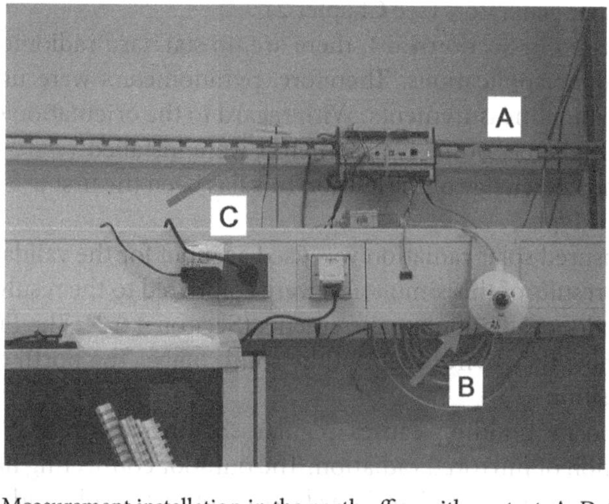

Figure 4.42 Measurement installation in the north office with context. A: Datalogger, B: Pyranometer *N3*, C: Room temperature sensor, behind wall temperature sensor.

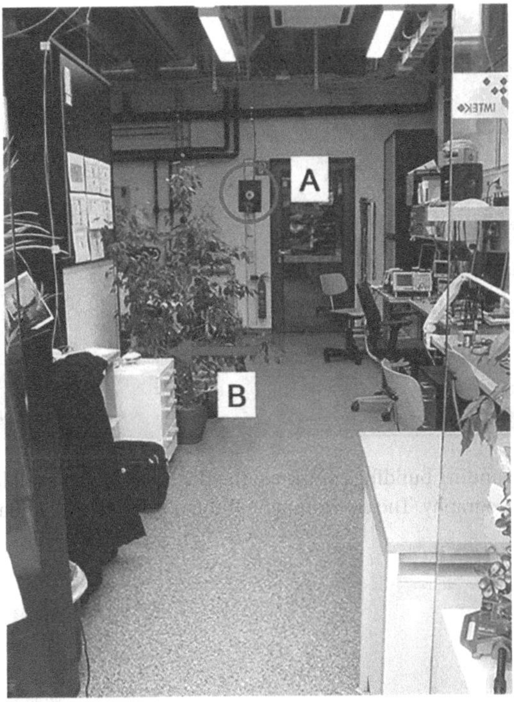

Figure 4.43 Environment of the installation location of the pyranometer *N1*(A) and *N2*(B) in the north office. Simulations were performed for the same locations.

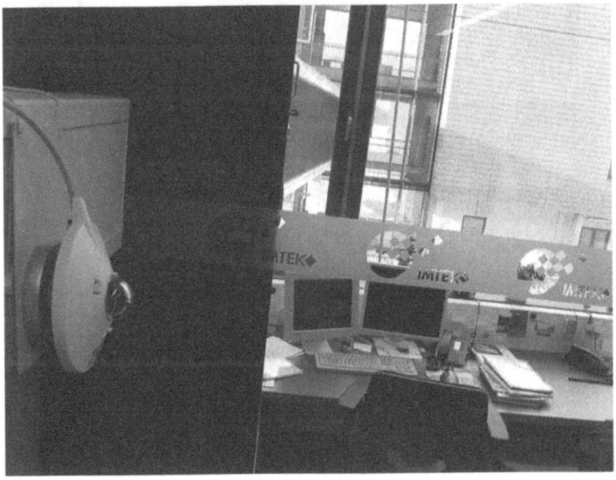

Figure 4.44 Vertical installation of the pyranometer *N3* in the north office. The simulation was performed with the same orientation.

Figure 4.45 Surrounding buildings of the south office viewed from soil level, hemispherical photography. The environment of the north office is similar.

Figure 4.46 Outdoor connection hallway between buildings, south side. A similar hallway leads to shading of the north window.

Figure 4.47 Installation of the radiation sensor, outdoor, north window. The sensor is oriented parallel to the window.

rooftop. Furthermore, the optically relevant parameters of the surrounding architecture would need to be measured for a validation measurement and integrated into the simulation model.

In order to avoid this tremendous effort and potentially resulting errors from the simulation of this complex environment, radiation sensors were mounted outdoors at the entry point of daylight in the simulation. For this type of sensor, too, so far no vertical installation has been reported so far.

Results and Discussion

Measurement data were recorded from 19-12-2008 to 06-01-2010. Missing data due to maintenance of around 8 days for the outdoor radiation measurement in the north office caused an uncertainty of 2%. The measurement data of the pyranometer measurements in the north office showed major errors and were extracted. This equals a portion of 6.5% of the measurement data. In the south office, the measurements of all radiation sensors were interrupted for 6 weeks with a resulting uncertainty of 11%, referred to the mean of the total interval. This data loss is due to the measurement effort of the evaluation of the new measurement procedure developed for indoor radiation in the same study (see Section 4.6), which was performed at the same time due to timing reasons of the overall study. Future application will not require this effort, so their outage times should be comparable to outdoor standard measurement setups.

Despite these limitations, the curve shapes and orders of magnitude from simulation and measurements are already comparable when using the Meteonorm data for global radiation. Figures 4.48 to 4.51 show the results of the measured radiation graphically; Tables 4.12 and 4.13 summarize the most important parameters for each measurement point. Due to the vertical installation of the outdoor radiation sensors, the measurement values for direct and diffuse radiation "direct" and "diffuse" are marked as modified and denote direct and diffuse solar radiation normal to the classical horizontal measurement orientation.

The following section discusses the influence of weather data. The measured radiation data of the reference data were then used as input data for the simulation of the models in Radiance. The dataflow is depicted in Figure 4.19.

4.6.3 Validating Simulation

In the final simulation, the measured radiation data were used directly as input data for the weather directly in Radiance. Missing data were interpolated

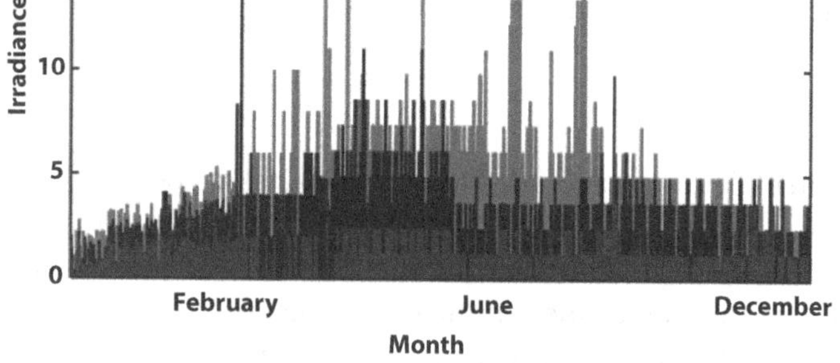

Figure 4.48 Measured indoor radiation, reference year 2009 in [W/m²], room with north window.

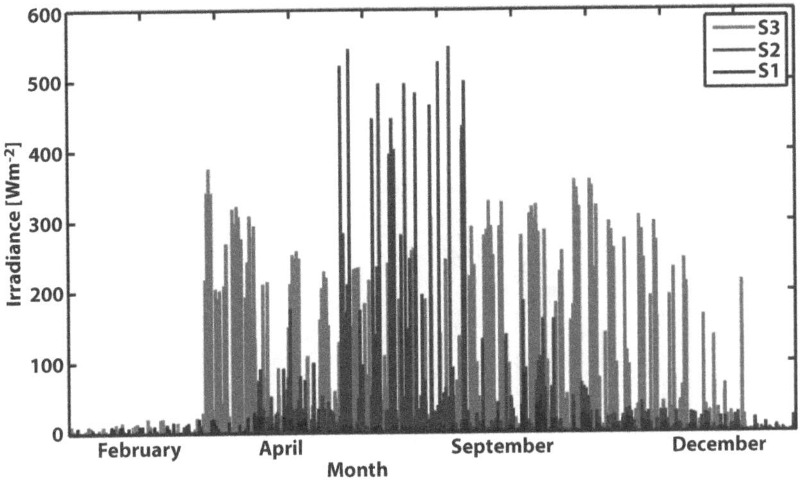

Figure 4.49 Measured indoor radiation, reference year 2009 in [W/m²], room with south window.

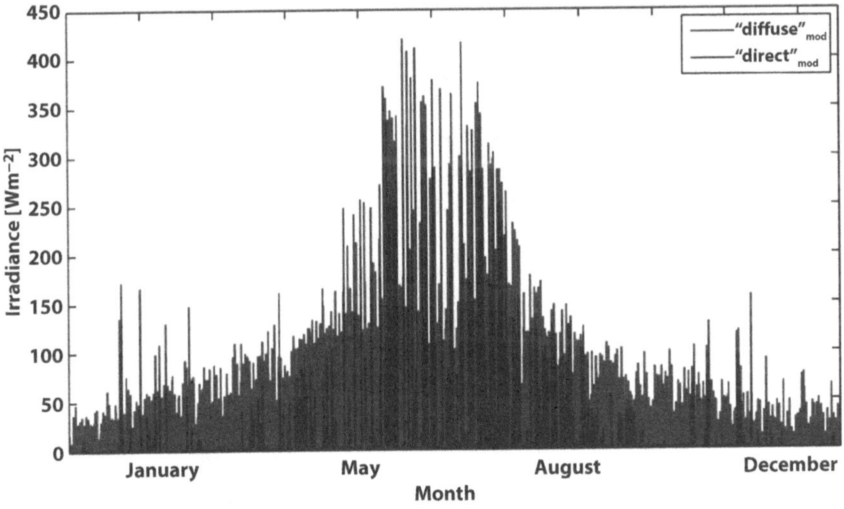

Figure 4.50 Measured direct and diffuse radiation in the reference year 2009 in [W/m²], room with north window, sensor oriented parallel to window.

for graphical depiction and in order to model functions, but were excluded from the evaluation. In accordance with the analytical modeling, the results for the 21st of June, the 23rd of March, the 23rd of September, and the 28th of December. The 28th December 2009 was chosen, as no artificial light were used on this day. For constructional reasons, the original data for the

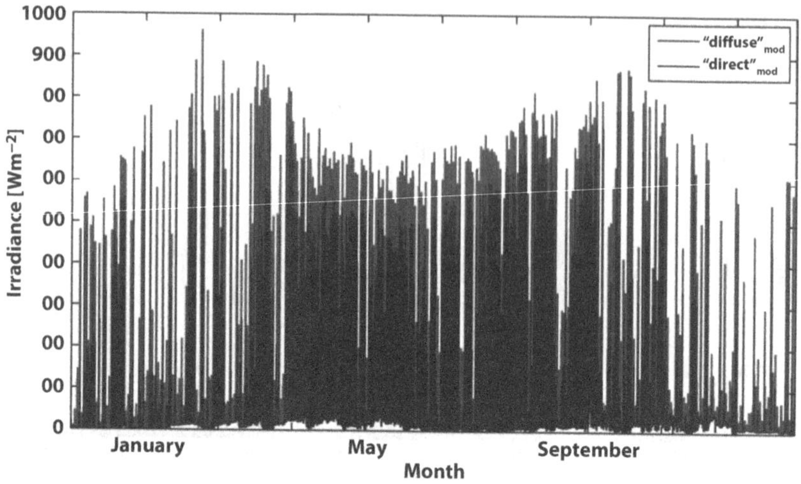

Figure 4.51 Measured direct and diffuse radiation in the reference year 2009 in [W/m^2], room with south window, sensor oriented parallel to window.

Figure 4.52 North office in *Radiance*, location *N1*.

transmissivity coefficient of the windows from 1996 were not available. Instead, data from similar window of similar technology at that time were taken. For the south windows, a g-value of a double-isolation window of T_n = 0.38 was assumed. Soiling and frames were assumed to reduce T_n to 0.2907 and t_n = 0.4867. The details are outlined in Section 4.5.7 and A.3).

Table 4.12 Characteristic values of measured indoor radiation in the reference year.

North	**Annual sum E_e**	$\hat{E}_e$	$\bar{E}_e = V$	σ
N1 in Wm²	33.433	9.77	0.32	0.55
N2 in Wm²	106.140	16.00	1.01	1.36
N3 in Wm²	142.730	22.09	1.30	1.69
South	**Annual sum E_e**	$\hat{E}_e$	$\bar{E}_e = V$	σ
S1 in Wm²	55.536	84.24	0.57	2.1
S2 in Wm²	338.460	548.23	3.45	18.1
S3 in Wm²	913.620	375.99	10.03	34.1

Table 4.13 Characteristic values of measured radiation in the reference year.

North	**Annual sum E_e**	$\hat{E}_e$	$\bar{E}_e = \mu$	σ
"Direct"$_{mod}$ in Wm²	208.370	384.24	1.98	11.75
"Diffuse"$_{mod}$ in Wm²	2.07e6	420.60	19.66	31.74
South	**Annual sum E_e**	$\hat{E}_e$	$\bar{E}_e = \mu$	σ
"Direct"$_{mod}$ in Wm²	4.87e6	837.73	49.72	122.02
"Diffuse $_{mod}$ in Wm²	7.68e6	962.63	78.35	150.53

Table 4.14 Measured and simulated in Radiance irradiation on significant days during the year, north office, without user model.

Sensor	$\bar{E}_{e\,June}$	$\hat{E}_{e\,June}$	$\bar{E}_{e\,Sept}$	$\hat{E}_{e\,Sept}$	$\bar{E}_{e\,Dec}$	$\hat{E}_{e\,Dec}$
Window	41.84	377.58	16.77	71.60	19.20	52.67
N1	0.37	1.22	0.59	2.44	0	0
$N1_S$	0.85	7.75	0.34	1.47	0.4	
N2	1.30	4.91	1.50	3.68	0.09	1.22
$N2_S$	0.62	5.52	0.25	1.05	0.28	
N3	1.69	4.88	1.52	4.88	0.33	2.44
$N3_S$	2.30	20.55	0.92	3.94	1.08	

Subscript "S" denotes a simulated value.

Table 4.15 Measured and simulated in *Radiance* irradiation on important dates during the year, south office, without user model.

Sensor	$\overline{E}_{e\,June}$	$\widehat{E}_{e\,June}$	$\overline{E}_{e\,Sept}$	$\widehat{E}_{e\,Sept}$	$\overline{E}_{e\,Dec}$	$\widehat{E}_{e\,Dec}$
Window	97.08	605.15	96.75	573.59	18.65	185.55
S1	0	0	0	0	0.07	1.22
$S1_S$	3.90	38.47	9.07	53.59	1.74	16.78
S2	0.41	3.66	0.49	2.44	1.17	9.76
$S2_S$	9.27	44.65	21.50	126.95	4.13	39.77
S3	3.14	14.63	2.33	10.94	7.14	69.30
$S3_S$	17.87	8.44	41.53	504	7.97	76.85

Subscript "S" denotes a simulated value.

The north window is a heat protection glass with a special EMALIT coating. A $T_n = 0.2$ with $t_n = 0.3785$ was assumed, including effects of soiling and frames. These relatively low values are in accordance with the level measured in the visible range. The measured transmission of both windows was $T_n = 0.58$ (north office) and $T_n = 0.63$ (south office) in the visible range. For today's technological standards with ($T_n > 0.75$) these values are in the lower range.

The Tables 4.14 and 4.15 show the irradiance measured at each investigated point on the north office, and in the south office, respectively. The results are compared to the simulation results obtained from Radiance as described in this chapter.

Table 4.16 summarizes the results obtained after including the user behavior model as outlined in the beginning of the chapter. For the north office, the simulation results in Radiance are tendentially too high. This is especially valid for the sensor *N2* oriented toward artificial light. In reality, artificial light was used for approximately 5 hours on this day, which was not yet included in this simulation. Similar to that, the simulated results for the overall office are significantly below the results for the measured irradiation. On this day, the glare protection was used for the entire day, but not included in the simulation model.

When taking into account that not all objects, such as the glare protection, were measured exactly, and the weaknesses of the measurement method discussed above for such a complex and vivid test environment, a simulation is a valid alternative to elaborate measurements.

The following section provides approaches to optimize measurements and simulations for indoor applications.

Table 4.16 Influence of user behavior on the example of measured and simulation irradiance. Sensors with an "S" denote simulation results, the others measured results. The simulation results for June were multiplied by a factor of 0.25 for glare protection. For the simulation results for September, the simulation results for the use of electric light were respected.

Sensor	$\hat{E}_{e\,June}$	$\hat{E}_{e\,Sept}$	$\hat{E}_{e\,Dec}$
Window	377.58	71.60	52.67
N1	1.22	2.44	0
$N1_S$	1.94	2.66	1.10
N2	4.91	3.68	1.22
$N2_S$	1.38	3.05	0.78
N3	4.88	4.88	2.44
$N3_S$	5.14	4.37	2.93

The measured values for irradiance and temperature throughout the year and the days indicated the use of artificial light and glare protection. On June 21, the measured irradiance reached a maximum value of about 378 Wm^{-2} at the window. The significantly lower values indoors indicate the use of glare protection. The agreement between measurement and simulation increases significantly when assuming a transmission of 25% of the incident radiation due to applied glare protection. With this approach, the results for the point *N3* only differ 2.5 to 3% from their joint mean value. A value of 25% for the transmission has also been implied in the simple, dynamic glare protection model in DAYSIM, which was also applied in the DAYSIM simulations of this study [27].

The example of the 23rd of September shows a similar strong influence of the use of artificial light. For this measurement day artificial light was used in addition to daylight. Adding the electric radiation simulated in Radiance to the simulation results of the daylight reduces the standard deviation to values as low as 0.6 with a deviation from the joint mean of about 5%.

The last studied day in Table 4.16 was 31-12-2009. On this day, no artificial light, and no or very little glare protection was used. Here, the measurement point *N2* showed the largest deviations, which in this simulation receives only indirect radiation.

To conclude, it is essential to account for the use of artificial light and glare protection in simulations. The achieved accordance between simulation and measurements when including the user behavior is a positive result for IPV design. The validation still includes some uncertainties, such as

Table 4.17 Comparison of measured and simulated irradiance for sensors in two rooms, artificial light, Radiance model.

Irradiance [Wm^{-2}]/Point	Radiance	Gossen[1]	Reference module[2]
N1	1.19	1.5	0.8
N2	2.00	7.0	2.1
N3	0.43	3.0	1.2
S1	2.88	3.0	1.3
S2	0.83	1.0	0.8
S3	0.12	0.0	0.3

[1]Gossen Mavolux digital, radiometric mode.
[2]Yxis IXOLAR High Efficiency Solar Bits, IXOB17-12xl, I_{sc} calibrated under 0.1, 1, and 10 Wm^{-2} in 4.53.

missing data for the exact window transmission or factors such as shading of sensor by office users. For this reason, a second, more elaborate measurement was performed under controlled conditions. The following Section discusses the results and the use of different measurement methods.

4.6.4 Comparison Measurement Methods under Controlled Conditions

The results of the simulation of artificial light in Radiance were compared to results obtained from measurements. Measurements were performed with a monocrystalline reference module, a pyranometer (Figure 4.53) and a luxmeter with radiometric function (Figure 4.54).

The reference module was calibrated for the installed and simulated radiation source (Figure 4.55). The radiation source of the measurement stand was characterized with a spectroradiometer.

The absolute value for the irradiance at the measurement area was measured with two CMP3 and CMP11 pyranometer each and with the short circuit current of highly efficient monocrystalline silicon solar cell for low irradiance [34]. For constant irradiance, a known short circuit current at known spectral irradiance $E_{e\lambda}$ following IEC 60904-3 Ed.2 (2008) [16] and known spectral sensitivity $S(\lambda)$, the m_{mf} was calculated in order to calculate the short circuit current under the spectral irradiance $E_{e\lambda}$ for the fluorescent light of a OSRAM Lumilux 18 W/840. In this measurement setup, the short circuit current of the applied highly efficient silicon module was measured. With this procedure the reference module for this study was obtained.

Figure 4.53 Measurement setup with CMP3 pyranometers. Measurements were recorded after reaching the thermal equilibrium with the environment. The shielded sensor served as control sensor. The same procedure was performed for the CMP11-pyranometers.

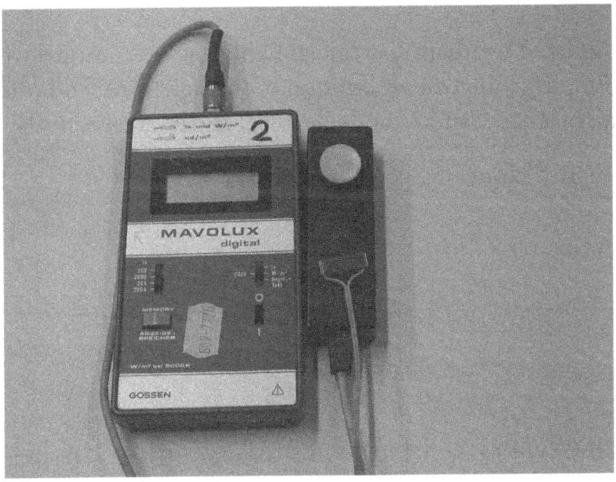

Figure 4.54 Reference module (right, on top of the luxmeter head) and luxmeter with radiometric function.

Reference module and simulation showed a high congruence for the north office, as can be seen in Table 4.17.

The model includes the IES data of the original illuminance, that is its emission characteristic, geometry, lamp ballast factor, and the power and spectral distribution of the installed fluorescent tube. The model in the south office used the installed luminants and a simple geometric model of the lamps. Table 4.18 compares the simulation results from Radiance for a simulation

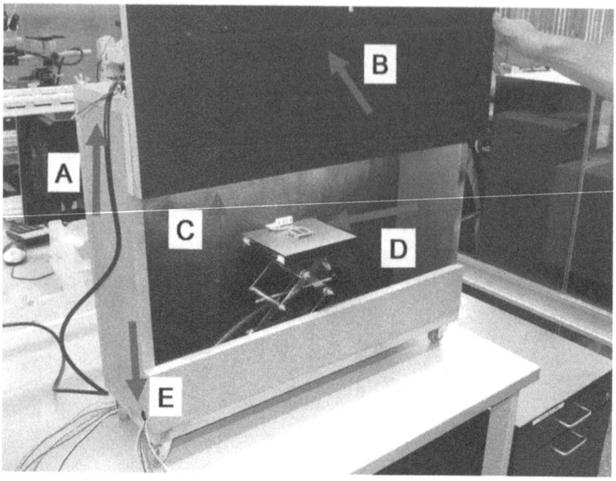

Figure 4.55 Measurement setup for the calibration of the reference module (see [1]). A: voltage regulation radiation source, B: radiation protection, C: top with radiation source, D: height-adjustable surface, E: measurement line.

Table 4.18 Simulated irradiance, artificial light, Radiance model with OSRAM Lumilux and simple geometric description of the light area, and original Trilux IES data for the geometry and lamp ballast factor 0.64, respectively.

Irradiance[Wm^{-2}]/Point	Geometry model	IES model
N1	2.42	1.30
N2	3.75	2.04
N3	1.94	0.94

of the north office with IES data, and a simple geometric model of the lamps. In accordance with the results from Section 4.6.3, the sensor with the most contributions of indirect radiation, thus the largest statistic proportion of the simulation, showed the largest deviation from the measurement results.

The installation point *N2* was simulated in Radiance. The validation measurement was performed with a reference module, a pyranometer, and a Gossen luxmeter with radiometric function. The value of the luxmeter exceeded the arithmetic total mean by 230%.

Due to the sensitivity of the pyranometer in the range of μV, thermal radiation and electromagnetic radiation from the environment are observable. Switching on and off the electric illumination was clearly visible in the results. Presumably, opening the doors influenced the results.

Table 4.19 Comparison of the results from measurements and simulation for the installation point *N2*.

Method	**Irradiance at *N2* [Wm^{-2}]**
Radiance[1]	2.04
Gossen Luxmeter[2]	7.00
CMP3 Pyranometer[3]	2.20
c-Si Modul YXIS[4]	2.13

[1] Simulation included the installed lamp with a OSRAM Lumilux Daylight 840W/18W.
[2] Radiometric mode.
[3] Setup as in Figure 4.54.
[4] Calibrated as in Figure 4.53 with OSRAM Lumilux Daylight 840W/18W.

The measurements of absolute values in order to validate the simulation thus required a more elaborate measurement setup. The results for N2 are displayed in Table 4.19.

The irradiance at *N2* was determined with a CMP3 pyranometer. The glass dome was isolated from thermal and optical radiation, in order to determine the measurement deviation in dependence from the room temperature and thermal radiation. The pyranometer was in thermal equilibrium with the environment at the beginning of the measurements. The output voltage was measured with two Agilent 6134 multimeters. The measured irradiance of 2.2 Wm^{-2} differed 10% from the simulated irradiance in the detailed model. The results between Radiance, the thermoelectric measurements with two CMP3 pyranometers and a monocrystalline module differed 4%.

4.7 Discussion and Recommendation

The existing commercially available measurements methods are only of limited applicability for IPV aplications. Most radiometric methods are optimized for an irradiance beginning at roughly 50 Wm^{-2} (see [30, 31]). On the other hand, measurement instruments for indoor use are mostly only available as photometric devices, thus rating an environment for its usability by the human eye based on their integrated $V(\lambda)$ filter. As a result, a portion of photovoltaically usable radiation is not detected. In addition to that, luxmeters are optimized to the spectrum of incandescent bulbs.

A deviation of 230% from the total mean was attained by the luxmeter in radiometric mode when measuring the installation point *N2* under fluorescent light of daylight spectrum. The deviation from the alternative, purely radiometric methods was 4%. The maximum deviation among all methods was 8%.

Due to their thermoelectric measurement principle, pyranometers are sensitive the smallest changes in their environment. While this principle enables the detection of signals as low as 0.1.1 Wm^{-2}, it also responds to thermal signal variations from the surroundings. Already an air draft at the sensor has destabilized the signal, which is only a few microvolts per Wm^{-2}. This is similarly true for electromagnetic disturbances, which affect the signal. For this reason it was necessary to build a stable environment, as it can be found in an unused temperature-controlled room, in order to prevent any unwanted thermal compensation current. The pyranometer needs to be at thermal equilibrium with the environment at the beginning of the measurements. This process took a few hours for this study. The required effort thus makes pyranometers in their current setting as unsuitable for mobile measurements in IPV applications. The broad range of spectral sensitivity is a plausible explanation of why the measured value is the highest obtained with radiometric methods.

Due to the signal-to-noise ratio and the resulting effort and measurement conditions outlined above, pyranometers were considered to be unsuited for most IPV applications at the time of this study. A valid application is the measurement of stationary installed light sources for constant environmental conditions.

The demonstrated method used, to calibrate a solar cells using a spectroradiometer and a monochromator specifically for the radiation source of interest, is the most effective and reliable method for most applications at the time of writing. Until IPV has gained a certain maturity, few commercial measurement instruments for this application field are to be expected. Thus, the most reliable methods for IPV designers is to use exactly the same material for detector and IPV generator and to calibrate the reference measurement cell to the dominating spectrum of the application. As soon as technologies for IPV have been established, the required spectral range of the measurement sensitivity will be known, and IPV measurement devices based on photodiodes can be developed. A more detailed discussion of materials best suited for IPV is provided in Chapter 3.

4.8 Conclusion and Outlook

For the first time, the work presented in this chapter systematically investigated orders of magnitude and typical irradiance levels for radiation in buildings on a radiometric base. This is essential for a reliable use of available optical radiation by photovoltaic generators.

Measured and simulated radiation were highly in accordance with their characteristic values and annual profile, although for the validation of the simulation of the dynamic contribution to radiation some real parameters remained unknown due to construction reasons. For constant electric lighting, an alignment up to 10% was achieved.

The validation showed the significance of the user behavior for the use of glare protection and artificial light. Both are implemented in DAYSIM with sufficient variability. Similar to the measurement devices, the next step for future work for IPV is to optimize the simulation possibilities in programs such as DAYSIM and Radiance for photovoltaic applications. Besides program extensions of the simulation evaluation, which is currently limited to daylight aspects, an extension of the modeled spectral distribution would especially be required. It would also be helpful to include an optional filter for wavelength intervals in order to simulate different photovoltaic materials.

In the field, relevant measurement methods include indoor radiation measurements with handheld devices and stationary long-term measurements and simulations. This chapter investigated the possibilities and limits of the current available solutions such as pyranometers, luxmeters, and the ray tracing programs Radiance and DAYSIM. The setups and results for experimental evaluation were presented. As a result, the indoor usage of pyranometers requires a stationary environment, that is constant room conditions including no temperature variations from air draft of other sources due to their low signal-to-noise ratio. If this is provided, pyranometers can be used for radiometric indoor measurements. As most applications are more dynamic, semiconductor-based sensors are currently the most important radiometric instruments for IPV applications. As most commercial products are designed for photometric applications, currently the recommended approach is to cast a promising photovoltaic material and to produce a sensor of the same material with a linearity and sensitivity optimized for the radiation level. This approach has been demonstrated for the calibration of a measurement set up for IPV module characterization [11, 35]. Chapter 3 includes theoretically ideal materials for IPV and measured indoor efficiencies with radiometric methods.

4.8.1 Autarky Factors

For the energy supply of off-grid households, the autarky factor A_F is set to A_F = 3 with respect to typical variations of the available solar radiation. In buildings, there are also variations of the energy from artificial light. Most often, the longest period without electric light depends on the presence of users. In the case of a public building, such as the one investigated in this chapter, the longest period without users depends on the type of institution. For a hospital, it might last 0 days, whereas for a school it might reach several weeks. For office buildings not operating on weekends, the period typically is 2 days, with more days added for buildings with plant holidays. The sensors studied in this chapter, which were oriented mostly toward an electric light, received a contribution of about 50% of their total received radiation from the electric light. For these types of installations, the A_F needs to be adjusted accordingly.

4.9 Acknowledgements

Part of this work has been performed while working on a PhD thesis at the Albert Ludwig University of Freiburg, Germany, in the Graduate School of Micro Energy Harvesting GRK 1322-1. The author is most grateful for the support from the university and dedicated researchers and industrial collaborators from all over the world during that work, and for the opportunity to successfully work on a PhD. The author also highly acknowledges the opportunity to publish this study in a revised and translated form, thus making the full content of this work available to a worldwide audience for the first time.

4.10 Symbols and Abbreviations

Symbol	Meaning	Used size, unit
a_Z	azimuth angle	°
A	area	m^2
A_F	autarky factor	1
AM	air mass factor	1
B(x)	radiosity	1
c_E	extinction coefficient	1
C	concentration factor	1
DC	daylight coefficient	1
E	intrinsic radiation	Wm^{-2}

(*Continued*)

Symbol	Meaning	Used size, unit
E_0	extraterrestrial solar constant	Wm^{-2}
E_e	irradiance	Wm^{-2}
$E_{e\lambda}$	spectral irradiance	$Wm^{-2}nm^{-1}$
$E_{e\lambda kal}$	spectral irradiance reference light source	$Wm^{-2}nm^{-1}$
E_{ind}	indirect irradiance	Wm^{-2}
$E_{p\lambda}$	spectral photon flow density	$m^{-2}s^{-1}nm^{-1}$
E_V	illuminance	Lux
$\overline{E_V}$	mean measured Illuminance	Lux
f_r	surface function	
F_{ij}	form factor	1
i	area	m^2
l	intensity	Wm^{-2}
I	current	A
l_0	saturation current	A
l_{ph}	photo current	A
l_{sc}	short circuit current	A
j	area	m^2
l	path through the atmosphere of the Earth	
l_0	thickness of the atmosphere of the Earth	
L_e	instrinsic radiation of the object	W
L_i	incident light	W
L_{ind}	indirect light	W
L_r	received light	W
m_{mf}	mismatch factor	1
n	refraction index	1
n	frequency	1
N	number	1
N_T	day of year	1

(*Continued*)

Symbol	Meaning	Used size, unit
P	power	W
P_E	received power	W
P_{Ein}	incoming power	W
P_L	radiation power fluorescent lamp	W
P_{sol}	solar radiation power	W
$r_{1,2}$radius/ distance	m	
r_s	mean radius sun	m
r_e	mean radius Earth orbit	m
s	standard deviation	1
s_N	inclination angle	°
S	Raumwinkelsegment	
S(λ)	spectral sensivity	A/W
S(λ)*cal*	spectral sensitivity of reference module	A/W
T	temperature	K
TN	transmission coefficient	1
V(λ)	human spectral sensitivity function	
X_j	random variable	
Y_k	random variable	
α	absorption coefficient	
α_{ges}	total angle	°
δ	declination	°
Θ	incident angle	°
λ	wavelength	m
Λ	longitude	°
Λ_0	reference meridian	°
φ	latitude	°
φ_i	photon flow density	$Js^{-1}m^{-2} = W$
Φ	angle between x and x’	°

(Continued)

Symbol	Meaning	Used size, unit
Φ_p	photon flow	s^{-1}
$\Phi_{p\lambda}$	spectral photon flow	$s^{-1}nm^{-1}$
ρ_r	reflection coefficient	
σ	standard deviation	
T_t	transmission coefficient	1
ω	hour angle	°
ω_f	angular frequency	s^{-1}
Ω	solid angle	°

4.11 Constants

Symbol	Meaning	Used size, unit
c_o	Vacuum speed of light	$2.997925*10^8\ ms^{-1}$
e	Euler number	2.7183
h	Planckian constant	$6.6256*10^{-34}$ Js
k_B	Boltzmann constant	$1.3805*10^{-23}\ JK^{-1}$
k_m	maximum of the photometric radiation equivalent	$683\ ImW^{-1}$
q	elementary charge	$1.6022*10^{-19}$ C
r_s	mean solar radius	$6.96*10^8$ m
R_U	mean distance Earth-sun	$1.496*10^{11}$ m
σ_{SB}	Stefan-Boltzmann constant	$5.67*10^{-8}\ Wm^{-2}K^{-4}$

4.12 Abbreviations

Abbreviation	Meaning
AM	air mass factor
CAD	computer-aided design
CIE	Commission Internationale de l'Eclairage

IEC	International Electrotechnical Commission
Ies	Illuminating Engineering Society
LED	light-emitting diode
STC	standard test conditions

Appendix

A.1 Luxmeter Accuracy

Table 4.20 summarizes the total error of all luxmeters discussed in this chapter. The standard deviation s was calculated based on the arithmetic mean $\bar{E}$ of al n measurements E_i:

$$s = \sqrt{\frac{1}{n-1}\sum_{i=1}^{n}(E_i - \bar{E})^2}. \tag{4.34}$$

Table 4.20 Measured illuminance for each luxmeter. The first error refers to the calculated statistical error (standard deviation *s*). The accuracy of luxmeter is known for operation under a *standard light source A* and listed as second error. For other light sources, a reference measurement might be required in order to measure the accuracy.

Illuminance E_V in Lux/Light source	**PeaktTech[1]**	**Gossen[2]**	**Meterman[3]**	**Minolta[4]**
Incandescent bulb CIE Class A OSRAM 40W	2352.8 ± 1.1 ± 94.1	2270.0 ± 1.4 ± 68.1	2233.3 ± 3.0 ± 66.9	2201.6 ± 1.4 ± 44.0
OSRAM Lu-milux Daylight 840/18W	3452.3 ± 0.2	4370.0 ± 0.0	3098.4 ± 5.6	4506.4 ± 1.4
LED Blue	152.9 ± 0.2	25.10 ± 0.0	101.3 ± 1.0	20.6 ± 0.2

[1] PeakTech 5030 digital luxmeter, accuracy 4%.
[2] Gossen Mavolux digital, accuracy 3%.
[3] Meterman LM631 LIGHT METER, accuracy 3%.
[4] Minolta T-1H, accuracy 2%.

A.2 Geometry and Materials in Radiance

Geometry and material parameters of an investigated scenario can be directly modeled as Radiance files. Alternatively, a geometry from common programs, such as Auto-CAD, can be used. In Radiance, rooms are built from walls, the so-called boxes, to which materials are attributed. The structure in Radiance is *genbox mat name xsiz ysiz zsiz [-r rad] [-b bev]*

where "mat" denotes the attributed material, "name" the designation of the particular "box", and "x", "y", and "z" the measure of length of the "box" in meters. With the options, "-r" and "-b", corners can be radiused or sloped.

The file geo.rad for the basic geometry for the investigated south room was written as follows:

"!genbox wallmat wallsouth 7.1 0.3 0.9/xform -t 0-0.3 0
"!gen box wallmat wallwest 0.3 5.3 3.28/xform -t -0.3 0 0
"!genbox wallmat wa I least 0.3 5.3 3.28/xform -t 7.1 0 0
"!genbox walldmat wall north 7.1 0.1 2.4/xform -t 0 5.3 0
"!genbox floormat floor 7.1 5.3 0.1/xform -t 0 0 -0.1
"!genbox ceilingmat ceiling 7.1 5.3 0.1/xform -t 0 0 3.28
*"!gensurf a_glassmat a_glass '7.1*s' '0' '2.38*t+0.9' 1 1*
*"!gensurf i_glassmat i_glass '7.1*s' '5.30' '0.88*t+2.4' 1 1*

The command "xform -t" translates the generated object along the reference coordinate system. The material parameters need to be declared before. This file is typically named as "mat.rad". Below is the example of the file "wandmat" in the simulation discussed in this chapter: *void plastic wandmat*

0
0
5 0.5 0.5 0.5 0 0

"Plastic" belongs to the classes predefined in Radiance, the following name specifies the exact meaning of the applied type. The first number designates the number of the following parameters. For the class "plastic," these are the values for red, green, blue, mirroring reflection and roughness. The measured total values for reflection were assumed to be equal for all colors and were implemented into the material files of the simulation. Furniture, plants and other room elements can be included in the model or included as submodel files.

A.3 Transmissivity Coefficient

A measured T_n already includes rays which were reflected within the material. Thus, the transmissivity-coefficient t_n needs to be determined. The transmissitivity-coefficient t_n for a refraction index $n = 1.52$ for glass is Radiance Tn

$$t_n = \frac{\sqrt{(1-r_n)^4 + 4(r_n \cdot T_n)^2 - (1-r_n)^2}}{2r_n^2 T_n} \tag{4.35}$$

where

$$r_n = \left(\frac{1-n}{1+n}\right)^2 \tag{4.36}$$

References

1. Hildebrandt, H., *Konzeption und Aufbau eines Teststandes zur Charakterisierung von Indoor-Solarzellen, Diploma Thesis*, p. 52, Freiburg, Albert-Ludwigs-University, Germany, 2009.
2. PCE, https://www.pce-instruments.com, accessed 2020-03-02.
3. Atandra, http://atandra.in/index.php/lux-meter/, accessed 2020-03-02.
4. EN 12464-1, Light and lighting. Lighting of work places. Version from 2003.
5. Department of Energy, *Energy Plus*, https://energyplus.net/, accessed 2020-03-02.
6. Metenorm - Global Solar Radiation Database, https://meteonorm.com/en/, accessed 2020-03-02.
7. Roth, W. *et al.*, *Photovoltaische Energieversorgung für Geräte im kleinen und mittleren Leistungsbereich, project report 0328689A*, Fraunhofer Institute Solare Energie Systems ISE, Freiburg, Germany, 1991.
8. Roth, W. and Steinhäuser, A., *Photovoltaische Energieversorgung von Geräten und Kleinsystemen, OTTI Technologie-Kolleg*, Fraun-hofer Institute Solare Energie Systems ISE, Freiburg, Germany, 1997.
9. Randall, J.F., *On the use of photovoltaic ambient energy sources for powering indoor electronic devices, PhD Thesis*, pp. 38–69, Lausanne, Ecole Polytechnique Federale de Lausanne, Switzerland, 2003.
10. Roundy, S. *et al.*, *Energy Scavenging for Wireless Sensor Networks: With Special Focus on Vibrations*, p. 26, Kluwer Academic Publishers, Norwell, MA, US, 2004.

11. Muller, M., *Energieautarke Mikroenergiesysteme am Beispiel von Photovoltaik in Gebäuden, PhD Thesis*, Osnabriieck, Der Andere Verlag, Germany, 2010.
12. Roth, W. and Steinhäser, A., *Photovoltaische Energieversorgung von Geräten und Kleinsystemen, OTTI Technologie-Kolleg*, p. 60, Fraun-hofer Institute Solare Energie Systems ISE, Freiburg, Germany, 1997.
13. Roth, W. *et al.*, *Photovoltaische Energieversorgung für Geräte im kleinen und mittleren Leistungsbereich, Project Report 0328689A*, pp. 3–17, Fraunhofer Institute Solare Energie Systems ISE, Freiburg, Germany, 1991.
14. Larson, G.W. and Shakespeare, R.A., *Rendering With Radiance: The Art and Science of Lighting Visualization*, Morgan Kaufmann Publishers, San Francisco, USA, 1998.
15. Walkenhorst, O. *et al.*, Dynamic annual daylight simulations based on one-hour and one-minute means of irradiance data. *Sol. Energy*, 72, 5, 385–395, 2002.
16. Standard Test Conditions as in IEC 60904-3 (2019). This study and its measurement setups was based on version IEC 60904-3, Ed. 2, 2008.
17. National Renewable Energy Laboratory, https://www.nrel.gov/grid/solar-resource/solar-glossary.html, accessed 2020-03-02.
18. Duffie, J.A. and Beckman, W.A., *Sonnenenergie: Thermische Prozesse*, 1. edition, pp. 25–37, Udo Pfriemer Verlag, Mu-nic, 1976.
19. Spencer, J.W., Fourier series representation of the position of the Sun. *Search*, 2, 5, 172, 1971.
20. Wagner, A., *Photovoltaik Engineering: Handbuch für Planung, Entwicklung und Anwendung*, 2nd Ed, p. 112, Springer, Heidelberg, Germany, 2006.
21. Hottel, H.C. and Sarofim, A.F., *Radiative Transfer*, p. 79 ff, McGraw-Hill, New York, USA, 1967.
22. Kaltschmitt, M. *et al.*, *Erneuerbare Energie - Systemtechnik, Wirtschaftlichkeit, Umweltaspekte*, p. 53, Springer-Verlag, Heidelberg, Germany, 2006.
23. Perez, R. *et al.*, Modeling daylight availability and irradiance components from direct and global irradiance. *Sol. Energy*, 44, 5, 271–289, 1990.
24. Feynman, R.P. *et al.*, *Feynman Vorlesungen über Physik. Band IMechanik, Strahlung, Wärme*, 4. Ed., p. 366, Oldenbourg, München, Germany, 2007.
25. Glassner, A. (Ed.), *An Introduction to Ray Tracing*, vol. II, p. 873, Academic Press, London, UK, 1989.
26. Larson, G.W. and Shakespeare, R.A., *Rendering With Radiance: The Art and Science of Lighting Visualization*, p. 524, Morgan Kaufmann Publishers, San Francisco, USA, 1998.
27. Reinhard, C., *Daylighting Handbook II: Daylight Simulations*, Building Technology Press, Cambridge, MA, USA, 2018.
28. Metenorm - Global Solar Radiation Database, https://meteonorm.com/en/, accessed 2020-03-02.
29. Hunt, D.R.G., The use of artificial lighting in relation to daylight levels and occupancy. *Build. Env.*, 14, 1, 21–33, 1979.

30. Solar irradiance sensors, Ingenieurbtüro Mencke und Tegtmeyer GmbH, https://www.imt-solar.com/about-us/, accessed 2020-03-02.
31. Kipp & Zonen pyranometer CMP3, CMP11, Signal Amplifier, https://www.kippzonen.com/, accessed 2020-03-02.
32. Indium Sensor X.ll, http://www.indiumsensor.de, accessed 2020-03-02.
33. Meier NT GmbH, *Heiko Meier Nachrichtentechnik*, datalogger ADL-MX https://www.meier-nt.de/en/, accessed 2020-03-02.
34. Glunz, S.W. *et al.*, High-efficiency silicon solar cells for low-illumination applications, in: *Conference Record of the Twenty-Ninth IEEE Photovoltaic Specialists Conference*, pp. 450–453, 2002.
35. Freunek, M. *et al.*, Maximum efficiencies of indoor photovoltaic devices. *IEEE J. Photovoltaics*, 3, 59–64, 2012.

5

Characterization and Power Measurement of IPV Cells

Stefan Winter

Department 4.5 Applied Radiometry, Physikalisch-Technische Bundesanstalt, Braunschweig, Germany

Abstract

For a reliable operation of energy autarkic Internet of Things (IoT) devices, a sufficient power supply is required. If the energy generation is realized with indoor photovoltaic (IPV) cells, their performance must be measured within the scope of quality management. However, the established methods for outdoor photovoltaics, as described in the IEC 60904 standards series, cannot be used one-to-one for this purpose. Instead, the special properties of the indoor application scenarios must be considered both in the standardized measurement conditions and in the measurement methods. As standardized spectra, since the spectrum of a white light LED is already standardized in photometry, the spectrum of a fluorescent lamp and the AM1.5 spectrum of natural sunlight can be used, each normalized to an illuminance of about 1000 lux, i.e., 100 times less than standard test conditions for outdoor measurements require. The corresponding light sources can be used for secondary calibration of IPV cells and the standard DSR method is very well suited for the high accuracy primary calibration and a comprehensive characterization of IPV cells.

Keywords: IPV, calibration, power measurement, indoor spectra, DSR, spectral responsivity, I-V characteristics, traceability

5.1 Features of IPV Compared to Outdoor PV

Energy harvesting devices based on solar cells must be optimized for the special requirements of their indoor application scenarios that differ significantly from those of solar photovoltaic energy systems in solar parks.

Email: Stefan.Winter@ptb.de

Monika Freunek Müller (ed.) Indoor Photovoltaics: Materials, Modeling and Applications, (115–132)

5.1.1 Irradiance

The most obvious difference between IPV and outdoor PV is the level of irradiance. In cloudless summer weather, a global irradiance of 1000 W/m^2 can be achieved if the solar module is oriented towards the sun and this value is defined in IEC 60904-3 [2]. This corresponds to an illuminance of roughly 100,000 lx (more precisely: 109.500 lx). In contrast, indoor lighting with artificial light sources only achieves illuminances of 500 lx to 1500 lx. These values are legally prescribed illuminances for work areas, e.g., office workplaces. However, since IPV devices are typically not located in brightly illuminated workspaces, they have in practice considerably less light for power generation. All in all, a hundredth to thousandth of the irradiance compared to a solar park must be expected. This aspect is very important, because typical solar cells, optimized for high irradiance and thus for high currents, must have a very good conductivity and often have more traps due to the higher doping, which in turn leads to photo augmentation at lower irradiance levels. While the traps are saturated at increasing irradiances and no longer play a role at 1000 W/m^2, the traps lead to an increased recombination rate of the electron-hole pairs at low irradiances and thus to significant efficiency losses. IPV cells must be optimized for low recombination rates, i.e., the minority carriers should have a high diffusion length. The success of the optimization must be checked for quality control.

5.1.2 Spectrum

Not only does the integral level of irradiance differ between indoor and outdoor applications, but also the spectrum of the light. The IEC 60904-3 standard defines the AM1.5 spectrum as the standard spectrum for outdoor applications [2]. It is defined for a surface perpendicular to the sun on which diffuse as well as direct sunlight is inclined toward. With the sun 42° above the horizon, the direct radiation effectively covers a distance through the Earth's atmosphere that is 1.5 times longer than when the sun is at its zenith. If the indoor area is illuminated through windows, then the resulting spectrum on the solar cell depends on many parameters, e.g., whether preferably reddish direct sunlight or bluish diffuse light of the sky enters, or the spectral transmission of the window and the spectral reflection of the ground. Therefore, the AM1.5 spectrum can be used as a rough approximation for one of the standardized spectra. In artificial lighting, the spectra of white-light LEDs and fluorescent lamps are the most common and therefore have already been standardized for lighting purposes by the

International Commission on Illumination (CIE) and can be taken from CIE 015:2018 for use in IPV [7]. For reasons of efficiency, incandescent lamps are almost never used in factory buildings and are being used less and less in residential applications.

5.1.2.1 Consequences of the Different Spectra Regarding Efficiency

The choice of the correct spectrum is of considerable importance for the future application. Even if all three spectra make a workplace appear equally bright, the IPV cell can generate different electrical powers from them. The reason for this is, for example, that natural light still has a certain amount in the infrared spectral range that the eye cannot perceive, but that a silicon solar cell can still use. Conversely, the efficiency of a solar cell depends on the defined spectrum. The less radiation is in wavelength ranges that the solar cell cannot use anyway, e.g., in the infrared above 1200 nm, the efficiency is better. Therefore, one and the same solar cell can have an efficiency of 20% in sunlight, but 30% if an LED spectrum is used as reference spectrum. For a more detailed discussion, please see Chapter 3.

5.1.3 Incident Angle Distribution

Since the revision of IEC 60904-3 in 2019, the angular distribution of incident radiation has been precisely defined. For metrological reasons, it is assumed that the entire radiation hits the solar cell perpendicularly, which is also approximately the case in a solar park at noon. For indoor applications, however, there is no generally preferred angle and there is a high diffuse radiation component. For standardized application conditions, a completely diffuse radiation distribution would therefore be obvious, but the standardization process has just started.

5.1.4 Modulated Light Sources

While natural sunlight hits the solar cell continuously over time, artificial light sources are almost always modulated for efficiency reasons. The modulation frequency of fluorescent lamps is 100 Hz or 120 Hz, depending on the line frequency. For LED light sources, the modulation frequency is in the kHz range, depending on its power supply unit. In order to consider these effects metrologically, it is recommended to measure first with a scientific, non-modulated lamp and then to determine a possible time-dependent behavior by varying the modulation frequency.

5.1.5 Further Effects

The non-uniform illumination that is possible in the area of IPV and that will not be considered in the standardized measurement conditions provides a subordinate effect. This is because IPV cells are often tiny compared to PV modules (e.g., 1 mm^2 in comparison to 2 m^2). In addition, effects due to curved surfaces are considered.

5.1.6 Standardization

All these aspects must be considered in the design of the IPV cells as well as in the standardized test conditions. Standardized test conditions are necessary so that not every manufacturer of IPV cells uses test conditions optimized for their own cell type with the consequence that the efficiency data in the data sheets are no longer comparable.

Figure 5.1 shows three preliminary spectra from the ongoing standardization process within the IEC TC82 WG2. These are a) the global AM1.5 solar spectrum, b) a fluorescent lamp spectrum and c) an LED lamp

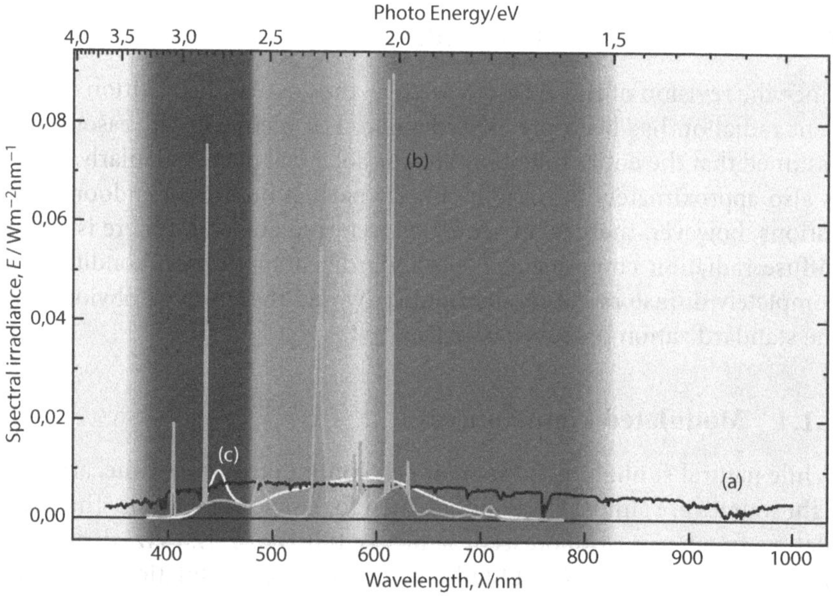

Figure 5.1 Diagram with (a) AM1.5 spectrum from IEC 60904-3, (b) fluorescent lamp spectrum of the narrow band lamp FL10 with a correlated color temperature (CCT) of 4000 K (cool white) and (c) LED spectrum (phosphor type LED-B3) with a CCT of 4100 K (cool white). All spectra are normalized to an illuminance of 500 lx. The spectra (b) and (c) are based on CIE 015:2018.

spectrum. The standardized irradiance is expected to be one hundredth to one thousandth of the solar irradiance. The temperature will be 25°C, as in the standard test conditions for outdoor applications.

5.2 Calibration Chain and Quality Management

For a reliable operation of energy autarkic Internet of Things devices, a sufficient power supply is required. If the energy generation is realized with Indoor PV cells, their performance must be measured within the scope of quality management. The quality management requires a proven and unbroken traceability chain by tracing the obtained measurement results back to the seven fundamental constants of nature like the speed of light, the Planck's constant or the elementary charge according to the International System of Units (SI). The measurements are done in several steps (metrological pyramid, see Figure 5.2a). The PV industry uses company internal measurements for production control. Their measurements are based on transfer standards (so-called "Golden Cells") from the production, which have previously been calibrated in accredited calibration laboratories. These laboratories use secondary calibration methods that require reference solar cells, e.g., in the World Photovoltaic Scale (WPVS) design (see Figure 5.2c). These WPVS-reference solar cells are calibrated against reference radiometers like electrical substitution radiometers or reference photodiodes (see Figure 5.2b) that where calibrated with the help of a high accuracy cryogenic radiometer, which is the national radiometric standard. Those measurements require a lot of effort and are performed by the National Metrology Institutes (NMIs), e.g., the Physikalisch-Technische Bundesanstalt (PTB), the NMI of Germany, that is the most active NMI in the area of reference cell calibrations for calibration and testing laboratories. The NMIs are responsible for the traceability to the fundamental constants of nature.

5.2.1 Basic Laboratory Measurement Methods for the Secondary Calibration of IPV Cells

The structure for determining the electrical power of IPV cells essentially consists of an optical and an electrical component. A light source with the above standardized properties generates the optical power that impinges on the IPV cell and is converted into electrical power by it. The maximum electrical power that can be generated is then determined with an electrical current-voltage characteristic measuring device. The quotient of electrical and optical power is the efficiency of the IPV cell. For dimensioning IoT

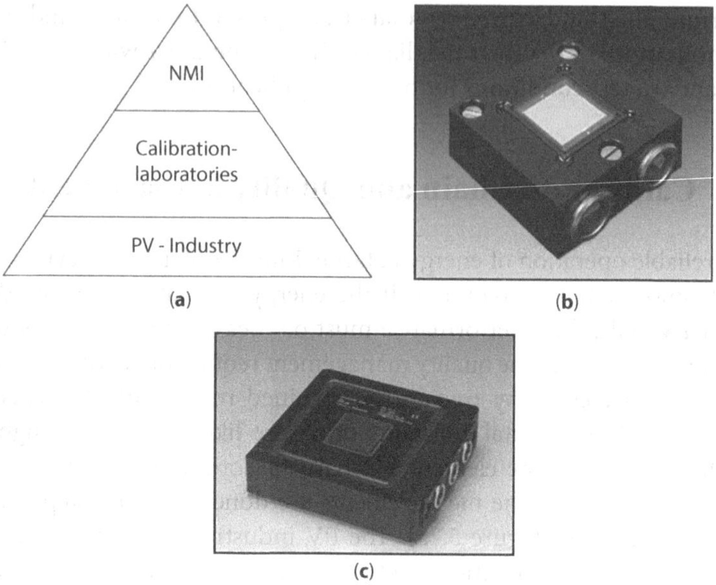

Figure 5.2 (a) The metrological pyramid with the NMIs at the top and the accredited calibration laboratories in the middle, which bring the unit to the PV industry. (b) Photo of a reference photodiode that is traceable to the national radiometric standard, a cryogenic radiometer. (c) Photo of a reference solar cell in WPVS design.

applications, it is also important to specify the electrical power that can be generated under standardized conditions.

Lamps, whose spectra come as close as possible to the spectra shown in Figure 5.1, should be used as light source. Since the standard uses spectra of typical commercial lamp types for artificial light sources, it is relatively easy to build a solar simulator, i.e., an indoor light simulator. However, consumer power supplies must not be used for the electrical supply, as they generate modulated radiation. Instead, laboratory power supplies must be used to generate constant light. In addition to the spectrum and stability, the third quality criterion of the indoor light simulators is the uniformity of the radiation field [5]. Even if the IPV cell is much smaller, the radiation field should have a uniformity of at least 2% in an area of 2×2 cm^2 to fulfill class A criteria. For the determination of the absolute irradiance, typically a primary calibrated reference solar cell of the size 2×2 cm^2 is used, for example, in the WPVS design.

To determine the maximum electrical power that can be generated by the optical power, for example, a 4-quadrant power supply unit is used to record the current-voltage characteristic curve. For each point of the characteristic curve, the product of the set voltage and the resulting current is calculated. Its maximum value is the maximum electrical power that can be generated.

If measurements are to be as accurate as possible, the deviations from the standardized ideal conditions must be taken into account, i.e., non-uniformities of the radiation field, deviations of the temperature from 25°C and deviations of the spectrum.

The easiest way to consider non-uniformity is to use the smaller object, usually the IPV cell, to measure its output signal at various points across the entire illuminated area covered by the larger object, usually the reference solar cell. This leads to a mean output signal that includes the non-uniformity of this illuminated area.

For temperature corrections, the temperature of the IPV cell must be measured with a temperature sensor. Alternatively, the temperature of the p-n junction can be determined according to the IEC 60904-5 standard [3]. The characteristic curve can be corrected using one of the three methods specified in the IEC 60891 standard. This standard defines procedures to be followed for temperature and irradiance corrections to the measured I-V (current-voltage) characteristics of photovoltaic devices. It also defines the procedures used to determine factors relevant for these corrections.

The most complex correction is when the spectrum of the indoor light simulator deviates from the standardized spectrum (spectral mismatch, [4]). The first step is to measure the spectrum, for example, with a calibrated and characterized array spectroradiometer. In order to determine the influence of the deviating spectrum, the relative spectral responsivity of the IPV cell must be determined. The procedure described in the next section for the primary calibration of IPV cells can be used in simplified form to do this. The spectral responsivity of the used reference solar cell can be taken from its calibration certificate. The spectral mismatch correction is performed with these spectral curves:

$$SMM = \frac{\int E_{\mathrm{ref}}(\lambda)s_{\mathrm{ref}}(\lambda)d\lambda \int E_{\mathrm{meas}}(\lambda)s_{\mathrm{IPV}}(\lambda)d\lambda}{\int E_{\mathrm{meas}}(\lambda)s_{\mathrm{ref}}(\lambda)d\lambda \int E_{\mathrm{ref}}(\lambda)s_{\mathrm{IPV}}(\lambda)d\lambda} \tag{5.1}$$

where

$E_{\mathrm{ref}}(\lambda)$ is the spectral irradiance of the reference spectrum,

$E_{\mathrm{meas}}(\lambda)$ is the measured spectral irradiance of the light simulator,

$s_{ref}(\lambda)$ is the spectral responsivity of the reference PV device at reference conditions,

$s_{IPV}(\lambda)$ is the spectral responsivity of the IPV device under testing at reference conditions.

All integrals shall be performed in the entire spectral range where the respective quantities are not zero.

If the light source is to weight all directions in the same way, as described in the paragraph "Incident angle distribution," an integrating sphere has to be used with the reference cell and the IPV cell placed at a port of the integrating sphere. At this port the light source may not directly be seen by the reference solar cell or the IPV cell. Direct light must be avoided by baffles.

5.3 Flexible and Precise Method for Comprehensive and Primary Calibration of IPV Devices

With the procedure described in the previous section, an IPV element can be calibrated very quickly. If a complete characterization, including the determination of the spectral responsivity, the spectral nonlinearity and the spectral temperature coefficient or even the spectral angle of incidence dependency is required, the method of differential spectral responsivity (DSR) is the best choice. Since no reference solar cell is required for this, but a highly accurate reference radiometer is used, this is a primary calibration procedure.

The goal of a solar cell calibration is the determination of the short circuit current under standardized test conditions (e.g., the STC AM1.5 spectrum according to IEC 60904-3, 1000 W/m^2 and cell temperature of 25°C [1, 2], or future test conditions for IPV as described before). This can be performed with integral measurements under a solar simulator or under natural sun. In both cases the spectrum of the source, especially with the difficulty of measuring its absorption lines, must be measured and taken into account [4], and this leads to additional uncertainty components [10]. The alternative is to determine the absolute spectral irradiance responsivity and to calculate the short circuit current using the formula

$$I_{STC} = \int_0^{\infty} s(\lambda) \cdot E_{\lambda,AM1.5}(\lambda) d\lambda \tag{5.2}$$

In this case, the exactly defined spectrum from the standard IEC 60904-3 is used directly; thus, there is absolutely no measurement uncertainty concerning the spectrum.

The main advantage for IPV application is that with the determination of the absolute spectral responsivity the current for any light source can be calculated. A further advantage is the possibility to measure the nonlinearity at any irradiances, that can be set by the bias lamps. The reference radiometer is only used without bias light; thus it does not have to be linear. The bias light for the solar device under testing is required to be set to the same working point as it would be in the use case, which means that the minority carriers should have the same diffusion length.

The spectral responsivity $s(\lambda, I_{STC})$ cannot be measured directly, because at standard test conditions with an irradiance of 1000 W/m^2 only the *differential* spectral responsivity $\tilde{s}(\lambda, I_{bias}(E))$ can be measured. This is realized with a two-beam method: In the presence of a sun-like bias light a modulated uniform monochromatic beam irradiates the solar cell and is measured with a lock-in amplifier. By integration over the wavelength, the AM1.5 weighted differential responsivity $\tilde{s}(I_{bias})$ is obtained:

$$\tilde{s}(I_{bias}) = \frac{\int_0^{\infty} \tilde{s}(\lambda, I_{bias}(E)) \cdot E_{\lambda,AM1.5}(\lambda) d\lambda}{\int_0^{\infty} E_{\lambda,AM1.5}(\lambda) d\lambda} \tag{5.3}$$

Using the formula

$$E = \int_0^{I_{STC}} \frac{1}{\tilde{s}(I_{bias})} dI_{bias}, \tag{5.4}$$

where I_{STC} is determined by iteratively evaluating the integral until the result equals 1000 W·m^{-2} [8]. Finally, the spectral responsivity for standard test conditions can be calculated [9]:

$$s(\lambda, E_{\mathrm{STC}}) = \frac{\int_0^{E_{\mathrm{STC}}} \tilde{s}(\lambda, E_{\mathrm{bias}}) \mathrm{d}E_{\mathrm{bias}}}{E_{\mathrm{STC}}} \tag{5.5}$$

The DSR method can be realized by different ways. They differ in the way the monochromatic light is generated.

5.3.1 Lamp-Based Facility

Coupling the light of a quartz halogen lamp and a high-pressure xenon arc lamp into a monochromator, monochromatic light in the wavelength range between 210 nm and 1900 nm can be generated [11]. While the irradiance of the bias light is about 1000 W/m^2, the optical power of the chopped light behind the monochromator is less than 100 μW at a spectral bandwidth of 5 nm to 10 nm. In principal, the bandwidth could be reduced by decreasing the width of the slits of the monochromator, but this would reduce the already low power even more. The same applies for the uniformity. A measurement with additional diffusers or at a larger distance would increase the uniformity and decrease the available monochromatic irradiance. To reduce this disadvantage some parts of the measurement, e.g., the measurements at different bias levels, are performed close to the monochromator to increase the signal to noise ratio, while other parts of the measurements are performed far behind the monochromator in a uniform monochromatic radiation field. But in this case the combination of both parts leads to additional uncertainties, especially if the solar cell is nonlinear. Nevertheless, for small solar cells the monochromatic irradiance is high enough and 2×2 cm^2 solar cells can be calibrated with an uncertainty of 0.5% (k = 2).

5.3.2 Laser-Based Facility

To solve the problems of the mentioned approaches a new type of facility for the primary calibration of reference solar cells can be used [12]. By using a tunable fs laser, the new setup (see Figure 5.3 and Figure 5.4) avoids the main problem of a lamp-based system: the low optical power of the monochromatic beam. The higher power of the new setup allows it to solve the subsequent problems of low monochromatic power, like:

- a bad signal-to-noise ratio for large solar cells at high bias radiation levels,
- the difficulty to reach a good uniformity,
- the high bandwidth of the monochromatic beam, and
- the interpolation error when merging absolute and relative measurements,

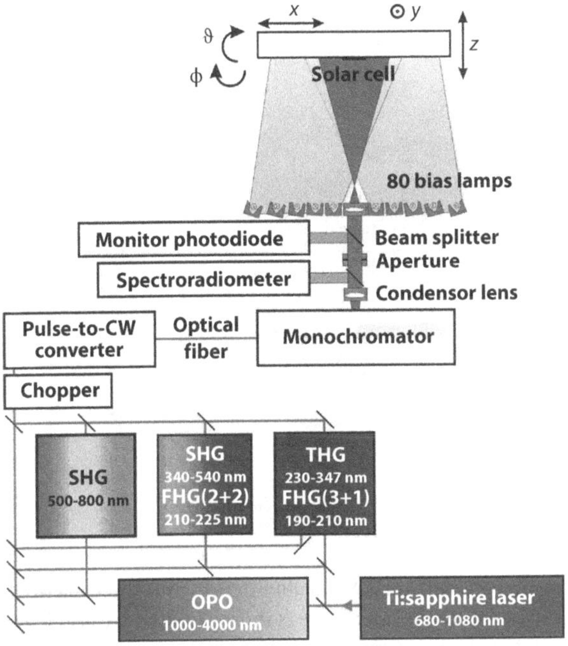

Figure 5.3 Schematic diagram of the Laser-DSR facility. In the lower half the laser setup is shown (see Figure 5.5). The beam passes a chopper before it hits an optical fiber (Pulse-to-CW converter). The fiber ends at the entrance slit of the monochromator. There the spectral bandwidth of the laser is reduced. The exit optics homogenize the modulated monochromatic radiation field on the solar cell. To get the same operating point as under natural sunlight, the solar cells are irradiated by a solar simulator consisting of 80 bias lamps. The spectrum of the bias lamps need not be identical to the reference spectrum, a similar color temperature would be enough.

because no compromise has to be accepted and all measurements at all wavelengths and all bias radiations are absolute measurements.

The full width at half maximum (FWHM) of the radiation, with values between 1 nm to 10 nm, is much lower than that of LEDs but at the same time is so high that no interferences appear, as it would be the case with CW lasers.

The principle of the Laser-DSR setup is identical to the standard DSR setup in many ways: A chopped uniform monochromatic beam and sun-like bias lamps irradiate the solar cell under test. The signal of the chopped beam is measured with a lock-in amplifier and is used to determine the absolute differential spectral responsivity and therewith the current under standard test conditions. The main difference of the new setup is the source of the monochromatic light. Instead of a quartz halogen lamp or a xenon lamp, the Laser-DSR setup uses a tunable laser that is coupled into the

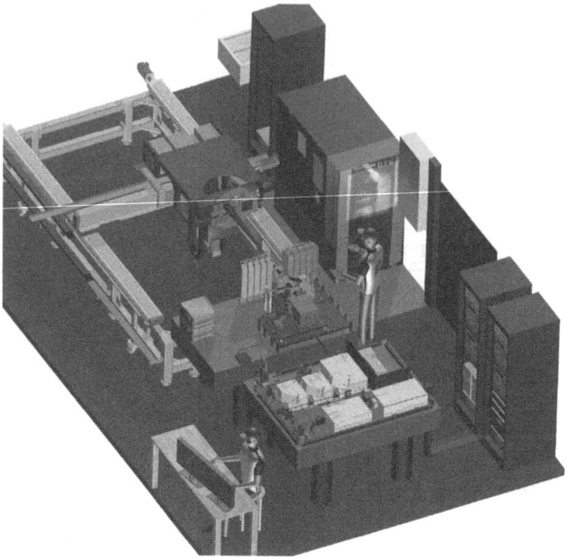

Figure 5.4 CAD drawing of the Laser-DSR facility showing the three major parts of the setup: On the lower right corner the laser setup is shown (yellow, see also Figure 5.5). The monochromator (red), based optics (green), as well as the bias lamps (orange) are located on the second table. The solar cells are mounted on the large x-y-z-table including the goniometer (see Figure 5.6).

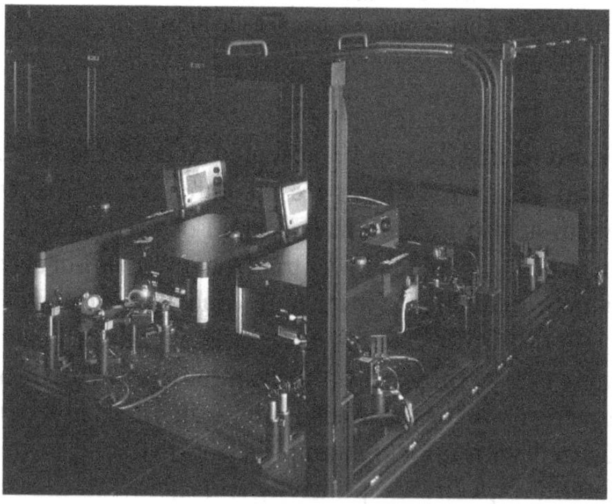

Figure 5.5 Photo of the complete laser setup with the beam routed through the SHG (400 nm). All beam paths are aligned to be coupled into the optical fiber (not shown here). During a measurement the laser setup is completely enclosed.

monochromator *via* a quartz fiber (see Figures 5.3 and 5.4). The laser beam starts at a hands-free, widely tunable mode locked Ti : Sapphire laser (Chameleon Ultra II) with a repetition rate of 80 MHz. The pulse duration is about 120 fs. Depending on the wavelength needed, the beam passes an optical parametric oscillator (OPO) and/or a second, third or fourth harmonic generator (SHG, THG, FHG; see Figures 5.3 and 5.4). Most of the laser components are purpose-built items that have a wider wavelength range, a higher efficiency or are better automated than the actual standard components. The beam routing is done by computer-controlled mirror mounts. This enables an automated wavelength selection from 210 nm up to 4000 nm (without automation even from 190 nm).

Before the laser beam is coupled into a fiber it passes through a neutral density filter wheel and a chopper. A monochromator reduces the spectral bandwidth and two lenses create a uniform monochromatic irradiation at the measuring plane. Up to 80 individually switchable bias lamps create a uniform bias irradiation of up to 10.000 W/m^2. The measurement plane can be tilted from -1° to 91° and rotated from -10° to 370° (see Figure 5.6). Thus, it can be used as a solar cell goniometer to irradiate the solar cell from any direction to measure a spectrally resolved angular dependence.

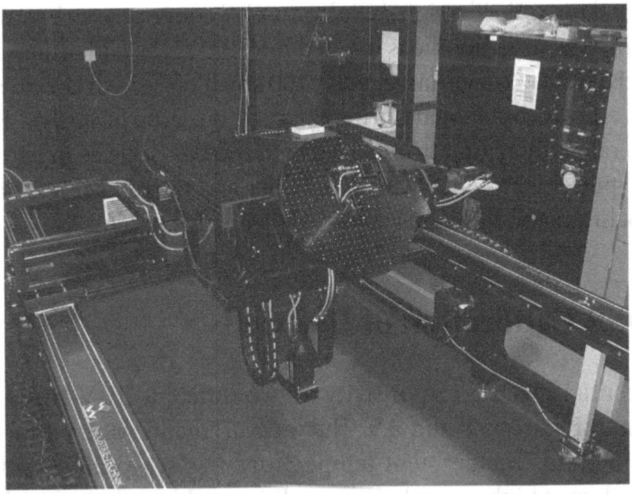

Figure 5.6 On the left-hand side of the photo the integrated solar cell goniometer can be seen. The quartz glass entrance window of the climate chamber is in the top right corner of the image. A mirror (not shown here) on the xyz table will reflect the monochromatic and the bias beam to a solar cell inside the climate chamber. The absolute spectral responsivity at 25°C will be measured on the xyz table and the relative change due to the temperature will be measured in the climate chamber.

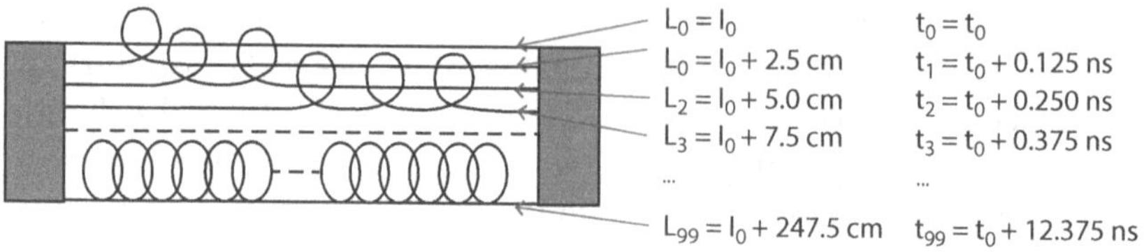

Figure 5.7 Schematic diagram of the Pulse-to-CW converter: A fiber bundle with 100 multimode fibers, each with an individual length.

The temperature of the solar cells is controlled by the combination of water cooling and Peltier elements.

A beam splitter reflects a small part of the monochromatic radiation behind the monochromator to a monitor photodiode. This enables a correction if the irradiance level of the monochromatic beam changes.

While the peak power of the femtosecond (fs) pulses within the laser system must be as high as possible to enable a wide wavelength range with the help of nonlinear effects (the efficiency of the wavelength conversion within nonlinear crystal increases with the irradiance), for the calibration itself constant radiation power is necessary to enable quantitative and reproducible results that are not biased by nonlinear saturation effects of the detector [13]. This challenge is solved by a special patented fiber design that smooths the pulsed signal to a CW-like signal (Pulse-to-CW converter, see Figure 5.7): The beam within the fiber bundle is divided into 100 paths with different lengths respectively. Instead of one large pulse, 100 small pulses exit the Pulse-to-CW converter, equally distributed during the cycle duration of 12.5 ns. Hence, the repetition rate is increased from 80 MHz to 8 GHz. Because in addition each of the 100 pulses is widened within the fiber and the following path through the monochromator, the signal at the exit is nearly constant.

5.4 DSR Calibration of IPV Cells

The differential spectral responsivity of two different III-V-based IPV cells are shown in Figures 5.8 and 5.9. The measurements were performed with the DSR method. The ambient temperature was set to 25°C, but the cells were heated up at higher irradiance conditions, although the bias lamps had reduced IR irradiance. But this increasing temperature reflects a realistic radiative heating at higher irradiances. The temperature was measured with a PT100. The sensor was pressed against the IPV chip from the rear. From the spectral responsivity curves, it can be seen that the chip has its responsivity mainly in the visible spectral range, i.e., in the wavelength range in

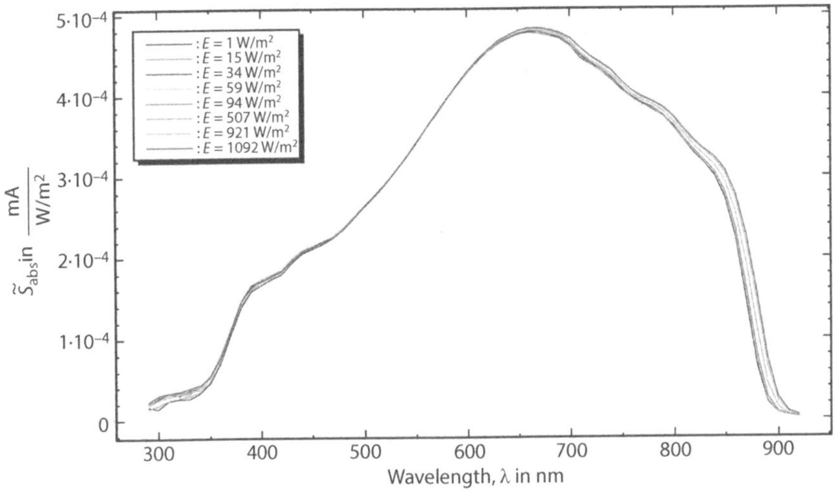

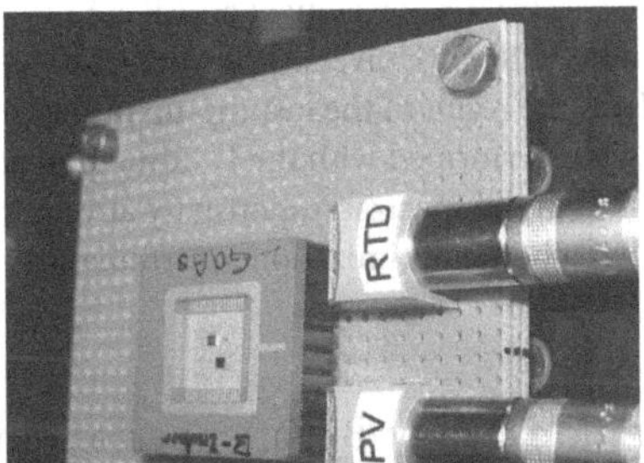

Figure 5.8 Absolute spectral responsivity of a GaAs IPV cell, produced by the research group of Jamie Phillips, University of Michigan, see Chapter 9. The nonlinearity in the UV and the IR is the result of the heating-up of the cell at high irradiance levels. The measurements were performed with the DSR method at PTB.

which modern, highly efficient light sources emit. Since the bandgap, and thus the open circuit voltage, is larger than for silicon, a higher efficiency can be achieved for these light sources. Due to the bandgap optimized for LEDs and fluorescence lamps, the thermalization losses are reduced.

5.4.1 Self-Referenced IV Characteristic

The last step of the calibration is the measurement of the IV characteristic of the IPV cell. This can be performed by using the above-described

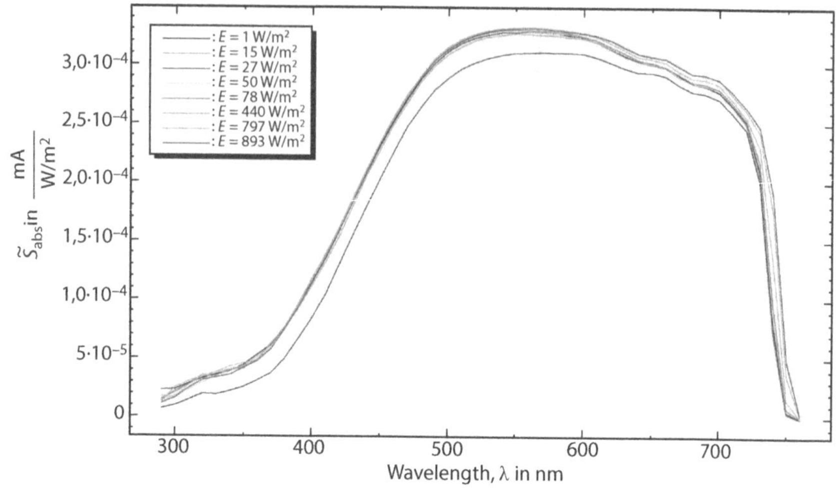

Figure 5.9 Absolute spectral responsivity of an AlGaAs IPV cell, produced by the research group of Jamie Phillips, University of Michigan, see Chapter 9. The nonlinearity in the UV and the IR is the result of the heating-up of the cell at high irradiance levels. The measurements were performed with the DSR method at PTB. In contrast to the GaAs IPV cell, there is a drop of the spectral responsivity at very low irradiances. The bandgap of the AlGaAs IPV has a better optimization for irradiance sources with only visible light.

secondary calibration facility. But after performing the DSR or Laser-DSR calibration, the most important parameter is already known: the current under standardized conditions. Thus, a further reference solar cell is not needed anymore for IV characteristic measurement. The DSR calibrated IPV cell itself is the reference cell for the irradiance. The irradiance of the light simulator is set so that the IPV cell generates the predetermined current under standardized conditions. With this setting the IV characteristic is measured. If the current cannot be set exactly to the previously determined value, the IV characteristic can be corrected according to one of the methods described in the IEC 60891 [6].

Acknowledgment

The project (Metro-PV, 19ENG01), leading to this publication has received funding from the EMPIR program co-financed by the Participating States and the European Union's Horizon 2020 research and innovation program.

References

1. *IEC International Standard 60904-1 Ed. 2: Photovoltaic devices - Part 1: Measurement of photovoltaic current-voltage characteristics*, Geneva 2006.
2. *IEC International Standard 60904-3 Ed. 4: Photovoltaic devices - Part 3: Measurement principles for terrestrial photovoltaic (PV) solar devices with reference spectral irradiance data*, Geneva 2019.
3. *IEC International Standard 60904-5 Ed. 2: Photovoltaic devices - Part 5: Determination of the equivalent cell temperature (ECT) of photovoltaic (PV) devices by the open-circuit voltage method*, Geneva 2011.
4. *IEC International Standard 60904-7 Ed. 4: Photovoltaic devices - Part 7: Computation of the spectral mismatch correction for measurements of photovoltaic devices*, Geneva 2019.
5. *IEC International Standard 60904-9 Ed. 2: Photovoltaic devices - Part 9: Solar simulator performance requirements*, Geneva, 2007.
6. *IEC International Standard 60891 Ed. 2: Photovoltaic devices - Procedures for temperature and irradiance corrections to measured I-V characteristics*, Geneva, 2009.
7. Carter, E.C., Schanda, J.D., *et al.*, Technical Report, *Colorimetry*, 4th Edition, 2018.
8. Metzdorf, J., Calibration of solar cells 1, The differential spectral responsivity method. *Appl. Optics*, 26, 1701, 1987.
9. Hinken, D., Kröger, I., Winter, S., Brendel, R., Bothe, K., Determining the spectral responsivity of solar cells under standard test conditions. *Meas. Sci. Technol.*, 30, 1–6, 2019. https://doi.org/10.1088/1361-6501/ab34ef.
10. Plag, F., Haas, F., Ramspeck, K., Nagel, H., Albert, H., Nevas, S., Sperfeld, P., Witt, F., Winter, S., Comprehensive analysis of a pulsed solar simulator to determine measurement uncertainty components, in: *29th European Photovoltaic Solar Energy Conference and Exhibition: Proceedings*, pp. 2435–2442, 2014.
11. Winter, S., Wittchen, T., Metzdorf, J., Primary reference cell calibration at the PTB based on an improved DSR facility, in: *Proc. 16th European Photovoltaic Solar Energy Conf*, vol. 4, H. Scherr, B. Mc/Velis, E. Palz, H.A. Ossenbrink, E. Dunlop, P. Helm (Eds.), James & James Science Publ., London, Glasgow, 2000.
12. Winter, S., Fey, T., Friedrich, D., Kröger, I., Von Volkmann, K., New laser-DSR facility at PTB: Concept for a next generation high accuracy primary calibration facility, in: *Proc. 26th European Photovoltaic Solar Energy Conference and Exhibition: The most inspiring platform for the global PV solar sector*, H. Ossenbrink, (ed.), and others, pp. 3463–3465, WIP Renewable Energies, München, 2011.
13. Fey, T., Kröger, I., Winter, S., Non-linearity effects of a detector due to pulsed radiation, in: *27th European Photovoltaic Solar Energy Conference and Exhibition*, pp. 1089–7690, WIP Renewable Energies, München, 2012.

6

Luminescent Solar Concentrators

Evert P.J. Merkx and Erik van der Kolk*

Delft University of Technology, Faculty of Applied Sciences, Section Luminescence Materials, Delft, The Netherlands

Abstract

Intended as an electricity generating replacement for windows, luminescent solar concentrators (LSCs) are at the borderline of indoor photovoltaics. In urban environments, space to install outdoor PV is limited to just the roof of a building, directly limiting the amount of energy that can be generated. LSCs aim to overcome this limitation by seamlessly integrating PV into the building envelope. This chapter explicates the luminescent phenomena and the working principle behind LSCs. Analytical and discretized techniques for simulating LSCs are presented and provided as downloadable examples. These same techniques show that an LSC can increase the performance of its PV frame with negligible dimming, even at non-perfect conversion efficiencies. This enhancement increases with window size. An overview of the optical properties of 28 state-of-the-art LSCs is presented. These LSCs are evaluated on performance for building-integrated purposes. Simulations show that LSCs with an optical efficiency of more than 2.8% are already possible for nontoxic quantum dot-based LSCs, without compromising on color rendering properties. Next to these state-of-the-art LSCs, a new development in the form of thulium-doped halides is highlighted. These halides are able to absorb the entire visible spectrum without coloration, and could be scaled to efficient LSCs of arbitrary sizes thanks to their lack of self-absorption. As IPV is an emerging application for LSCs, most of the chapter focuses on LSCs for window applications in order to outline the theoretical foundations. In addition to that, developments in LSCs for usage as indoor-only photovoltaic are illustrated with a fully worked-out example of an LSC as tabletop. Next to this IPV simulation, colorful designs that apply LSCs in other ways than just windows are highlighted.

Keywords: Building-integrated photovoltaics, fluorescent and luminescent materials, luminescence, computation methods, concentrators, solar energy

**Corresponding author*: e.vanderkolk@tudelft.nl

Monika Freunek Müller (ed.) Indoor Photovoltaics: Materials, Modeling and Applications, (133–188)

6.1 Introduction

Electricity-generating windows (PV windows) are literally at the borderline of conventional PV applications and IPV. Currently, the thousands of square kilometers of window area in the world are mainly[1] used to let sunlight pass and illuminate the inside of buildings. An active area of research is focused on using part of the incident solar spectrum to generate electricity with photovoltaic cells. Today's commercially available electricity-generating windows operate using thin and thereby semitransparent solar cells made out of conventional absorber materials like Si, CdTe, CIGS, or organics, laminated onto the glass window. Although impressive efficiencies have been achieved, they have not become the building-integrated (BI) industry standard because of the relatively complex multilayer technology used, which makes the windows too costly.

A luminescent solar concentrator (LSC) can be a cheap and much simpler single-layer alternative technology to realize electricity-generating windows. LSCs, as illustrated in Figure 6.1, were first studied in the 70s [1–3] as an alternative to lens-based concentrators. Being able to absorb both direct and diffuse light, LSCs do not need costly solar tracking systems to stay pointed towards the sun. Today, the value of LSCs is more in their potential as a building-integrated PV (BIPV) window [4, 5] or, more generally stated, as BIPV devices applied as or integrated with the transparent parts of the building's outer shell.

In this chapter, we will not consider applications of LSCs on non-transparent parts of a building. Applying an LSC on non-transparent areas causes the LSC to automatically compete with conventional (higher efficiency) mass-produced PV technologies, such as c-Si panels.

In this chapter we will give an overview of topics related to the development of an LSC intended for application as PV window. In Section 6.2, we will give a short introductory course on the various types of photoluminescent centers used in LSCs. In Section 6.3, we will give a general account of the principle of operation of an LSC by describing all separate optical and electrical conversion steps. Here we will follow the incident photons from light harvesting, to light trapping and transport and finally to photoelectric conversion, by giving a qualitative account. In Section 6.4, we will give a much more theoretical quantitative account of light transport through an LSC. This involves computations using analytical models as well as Monte Carlo simulations. These models can deal, for example, with the complex

---

[1]Further functionality of windows can be found in Chapter 4 of this book.

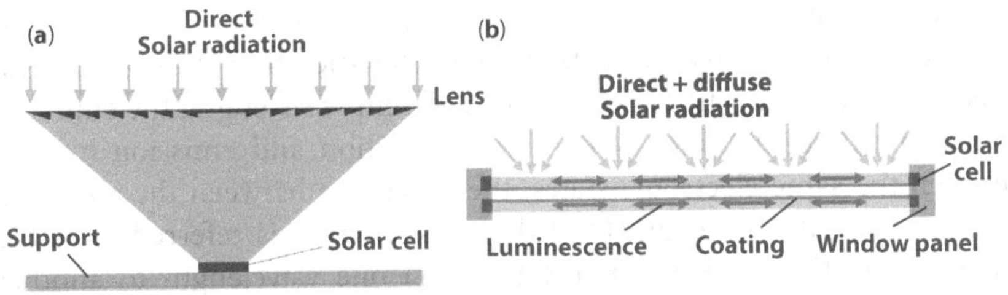

Figure 6.1 Traditional lens-based concentrator PV device (a) versus a PV-window concentrator device (b) based on the luminescent solar concentrator principle: Sunlight is absorbed by a coating on the window and is re-emitted by a process called luminescence in all directions. A significant fraction (75% to 90%) of the light is trapped by total internal reflection in the coating-window combination that acts as a light guide (like in a fiber optic cable) and directs it to the window edges where strip-shaped solar cells convert it into electric power.

behavior of self-absorption, or can be employed with the goal to obtain maximum possible power conversion efficiencies of an LSC applied as a window. We will translate the theoretical concepts to idealized material properties, that is, define the ultimate properties that a luminescent coating on glass should have to make an LSC PV-window with the highest possible efficiency. This section will logically be complemented with an overview of current state-of-the-art luminescent materials considered for LSCs, found in Section 6.5. In Section 6.6, we consider a recently emerged class of halide luminescent materials based on the element thulium in its 2+ valence state (Tm^{2+}), as these materials combine many favorable properties like a broad absorption range up to 900 nm and no self-absorption. Next to these more traditional uses of LSCs, a perspective on LSCs for full indoor use can be found in Section 6.7. Besides its energy conversion potential, we will also highlight some of the challenges ahead for realizing PV-windows and IPV devices. These challenges involve techniques to apply coatings on glass and ensuring material stability over long timespans. In the last section, Section 6.8, we conclude on the future potential of LSC PV applications.

6.2 A Crash Course in Luminescence

In order to understand the principle of luminescent solar concentrators, it is imperative to get some grip on what luminescence entails, and what machinery is behind it. If you are already familiar with the workings of luminescence, you can skip this section and continue reading on the working principles of LSCs in Section 6.3.

Luminescent centers, often also called luminophores, absorb light of various wavelengths based on their characteristic absorption spectrum. This absorbed light gets converted to a different wavelength: the emission of the luminescent center. When the absorption and emission happen between the same electronic states, the difference between the absorbed and the emitted wavelength (and therefore energy) is referred to as the Stokes shift. This conversion of light from one wavelength to another occurs with an efficiency known as the (photoluminescent) quantum yield η_{QY}, defined as

$$\eta_{QY} = \frac{\#\,\text{photons in}}{\#\,\text{photons out}} \tag{6.1}$$

Three different types of luminescent materials that are of particular relevance to LSCs–organic dyes, rare earth ions, and colloidal quantum dots–will be treated in the following sections. In these sections we will give a short overview of what the working principle of the luminophore in question is, together with an explanation of some frequently used terminology.

6.2.1 Luminescence in Organic Dyes

A large family of luminescent solar concentrators are based on luminescent dyes. This is mainly due to their high quantum yields and ease of handling due to their high solubility in waveguide matrices, like PMMA [6] and polycarbonates [7].

These dyes are π-conjugated organic molecules: molecules with a chain of carbon atoms having alternating single and double bonds, causing the p-orbitals of the carbon atoms to overlap. These dyes have a system of de-localized π-electrons-electrons where two orbital lobes from different atoms overlap-above and below the planar conjugated chain. Absorption in these dyes is caused by a promotion of the molecule from the singlet ground state S_0 to an excited state S_1 through absorption of a photon with sufficient energy. Subsequent luminescence is a result of relaxation of the molecule to the lowest S_1 excited state, followed by emission of a photon, leading the system to fall back to a higher S_0 state, which in turn is followed by non-radiative relaxation to the S_0 ground state. Figure 6.2 shows a summary of the processes involved in dye luminescence, as depicted in a Jablonski diagram. An excellent treatise of dye luminescence can be found in Ref. [8], with a more brief explanation of the underlying concepts in Ref. [9]. For dyes specific for LSCs, an overview can be found in the review by Debije and Verbunt [5]. An interesting property of the $S_0 \leftrightarrow S_1$ interplay is that the excitation spectrum mirrors the emission spectrum, as displayed for the popular LSC dye BASF Lumogen F Red 305 in Figure 6.3. In organic dyes, the shape of the absorption/emission spectra and Stokes shift

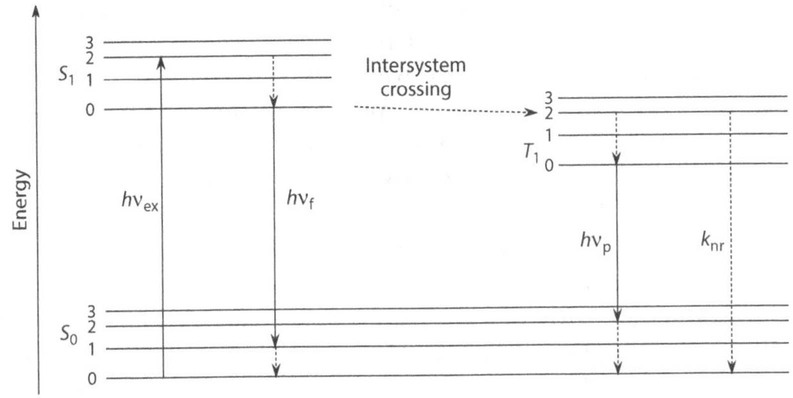

Figure 6.2 A simplified Jablonski diagram of luminescence processes in dyes. A photon with energy $h\nu_{ex}$ excites the system from the S_0 level. The system relaxes (non-radiatively, indicated by a dashed arrow) either (efficiently) from $S_1 \rightarrow S_0$ under emission of a photon with lower energy $h\nu_f$ (fluorescence), or the system moves to the excited triplet state T_1 through intersystem crossing. From the T_1 excited level, the system either relaxes through non-radiative relaxation k_{nr}, or through emission of a photon with energy $h\nu_p$ (phosphorescence). Non-radiative transitions to the ground state are not excluded for the excited singlet states, but more unlikely than for the triplet states.

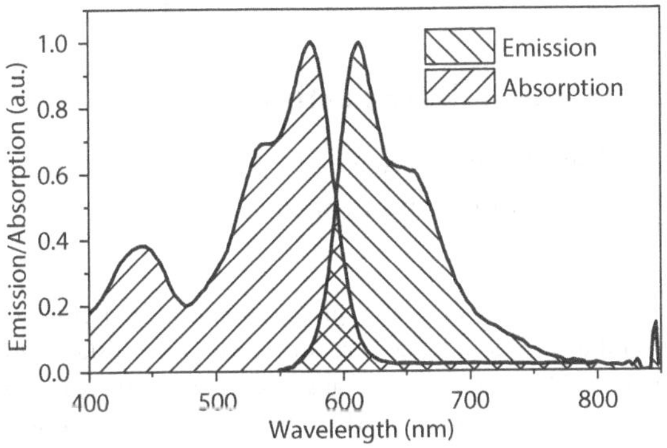

Figure 6.3 Normalized absorption and emission for BASF Lumogen F Red 305 [7]. The mirror image from absorption to higher lying singlet states and subsequent fluorescence is clearly visible. The cross-hatched area shows the overlap between absorption and emission. The area of this overlap should not be seen as a "self-absorption percentage," as both curves are normalized.

between them is mainly governed by the chain length and the number of π-electrons in the conjugated plane of the molecule.

The mirror-image of the absorption and emission is an immediate disadvantage of dyes for LSC use. As will be further explored in Section 6.3 on LSC behavior, having an overlap between absorption and emission can be detrimental to LSC performance when scaling up the LSC to dimensions spanning nearly a square meter. For IPV applications, at square centimeter sizes, the effect of this overlap will only manifest itself as having an impact on performance for high absorptions.[2] Absorption is only followed by emission if the quantum yield of the luminophore is sufficiently high. Therefore, if the dye is able to absorb its own emission, the LSCs overall performance will decrease when the LSC increases in size, since a larger LSC leads to more absorption taking place (and therefore also to absorption of its own emission).

Next to the limitation of self-absorption, dyes are also known to only have limited absorption ranges, causing a single dye to never be able to absorb the entire solar spectrum. A strategy to overcome this limitation is of course to combine multiple dyes. This can be done by either having multiple LSCs stacked on top of each other [5, 10, 11], akin to tandem solar cells, or by mixing all dyes together in a single waveguide [5, 12]. The absorption range of a dye can be altered by changing the conjugated chain. However, when increasing the chain length for moving the absorption to the infrared, the solubility of the dye in a matrix material decreases [13]. A lower energy between transitions also makes it easier for the dye to fall back non-radiatively from its excited state, leading to a decreased quantum yield when moving the emission to the near-infrared.

6.2.2 Luminescence in Rare Earth Ions

Inorganic hosts doped with rare earth ions[3] known as phosphors, are widely used in various optical applications, such as laser materials, (LED)

---

[2]The crucial part for the impact of overlapping absorption and emission is the product between distance and absorption. If light has to travel a long distance to the edges of the LSC, but has only a small chance of being absorbed per distance, the overall absorption during light transport will be the same as when the light only has to travel a short distance, but with a high chance of being absorbed per distance.

[3]Rare earths are the lanthanides, the elements with atomic numbers 57 to 71, from lanthanum (La) through lutetium (Lu), often depicted as its own row underneath the periodic table, combined with Scandium (Sc) and Yttrium (Y). "Rare" earth is a misnomer. While the rare earths in nature are difficult to separate, and are typically not found in clusters of pure elements, they are quite abundant in the Earth's crust [14].

phosphors, banknote security, and luminescent markers in the field of biomedics [15–17]. In the field of photovoltaics, various combinations of rare earth materials doped in a great variety of hosts are used as either up- or downconversion materials (though with $NaYF_4$ doped with Yb^{3+} and Er^{3+} as most prominent class for up-conversion) [18–21] to improve solar cell efficiency; and of course for use in LSCs, as will be elaborated on in this chapter.

A remarkable feature of the rare earth ions in compounds is their (partially) filled 4f shell. What makes these 4f electrons remarkable, is that because of the completely filled $5s^2 5p^6$ shells surrounding them, their intra-orbital transitions are well shielded from the influence of their environment. This means that, regardless of what host these ions are placed within, they will have (to a great degree) the same 4f energy levels [22, 23]. The intensity (and altogether presence) of transitions, and therefore emissions, from these f levels is dependent on the local symmetry and available phonon modes of the host.

Intra-f absorption is strongly forbidden by the parity selection rule, but not fully excluded due to mixing with opposite parity states [16, p. 26], and therefore absorption transitions between the f states are generally extremely weak. To become interesting for LSC applications, the much stronger (parity allowed) absorbing f-d transitions have to be considered. $4f \rightarrow 5d$ absorption is typically 10^2 to 10^4 times stronger than $4f \rightarrow 4f$ absorption.

The 5d orbitals of a rare earth are much more extended in space than the 4f orbitals, and hence don't have the same degree of shielding by the outer filled $5s^2$ and $5p^2$ shells. This means that the energy difference of the excited 5d level with the 4f ground state is strongly dependent on the host material. Therefore, it is possible to engineer the absorption range of a rare earth-based LSC by changing the host material. The energy of the 5d levels compared to that of a free ion, is mainly altered by the host in two ways: by ligand anion type, and by the structure of the host crystal, as illustrated in Figure 6.4.

If we imagine incorporating the cation in a host as surrounding the cation by a spherically symmetrical field with negative charge, screening the charge of our cation, we can observe the binding energy levels of the cation's states being lifted closer to the vacuum level. This effect is called the chemical shift and affects both the 4f and 5d levels. The intensity of this effect depends on the type of ligand anion and follows what is called the nephelauxetic[4] sequence. In order of raising the level closer to the vacuum level at 0 eV: F, oxides, Cl, Br, N, I, S, Se, Te. The shift in binding energy of

[4]Greek for "cloud expanding."

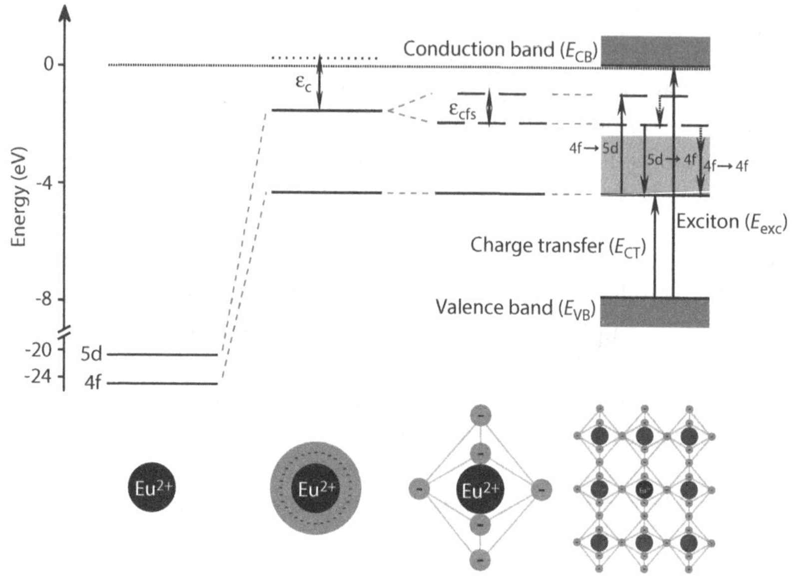

Figure 6.4 The development of the binding energy of an electron in a state of a divalent europium cation when it is incorporated in a host, relative to the vacuum level (at 0 eV). First, there is the free cation, with the electron's binding energy well below the vacuum level. Surrounding the cation with a spherically symmetrical field of negative ions leads to a shifting of the energy levels altogether-the chemical shift-and the centroid shift ε_c of the 5d level. Transitioning from this field to negative ligands splits the 5d levels,[5] yielding the crystal field splitting ε_{cfs}. Expanding the surrounding ligands to a crystal adds the valence and conduction bands. In this last frame the various types of possible transitions are also shown. The many 4f levels are indicated as a light gray area above the ground state.

the 5d levels is on average smaller than that of the 4f levels. To correct for this, the centroid shift ε_c is introduced. For ε_c the nephelauxetic sequence is also followed, with the difference between 4f and 5d being highest for fluoride ligands (i.e., low ε_c) and lowest for telluride ligands (i.e., high ε_c).

In reality, the host we place the cation in is not a spherically symmetrical field, but rather has a certain type of (local) structure to the position of the cation and its ligands: the coordinating polyhedron. The energies of the 5d levels are affected through their orbitals by the local structure of the host.[6]

[5]Because of spin-orbit (*LS*) coupling (not displayed in Figure 6.4), the 5d levels are already split, covering a small energy range. *LS* coupling is around half as intense as the splitting caused by the crystal field. In terms of overall influence to the energy w.r.t. the vacuum level a general rule-of-thumb is: chemical shift> $\varepsilon_c > \varepsilon_{cfs} >$ *LS* coupling.

[6]In principle, this also affects the 4f levels. However, the influence of the structure of the surrounding crystal on them is ~50× weaker than for the 5d levels.

This splits the 5d levels into different components at different energies depending on the coordinating polyhedron: the crystal field splitting. In practice, these splitting effects mean that rare earths can have absorption ranging from just UV absorption, to covering the entire visible spectrum. The splitting of the 5d levels can be observed by exciting the luminescent material at different wavelengths, illustrated for $CaBr_2:Tm^{2+}$ in Figure 6.5.

Next to excitation from a 4f → 5d transition, photoluminescence in rare earths can also be generated through the allowed charge-transfer excitation or host-lattice excitation (formation of an exciton), seen for $AlN:Eu^{3+},O^{2-}$ in Figure 6.5. In a charge-transfer excitation, an electron from the ligands L surrounding the rare earth cation is excited to the 4f level of the rare earth ($4f^nL^q \rightarrow 4f^{n+1}L^{q-1}$). Following this transition, the system relaxes back, but leaves an excited $4f^n$ state behind. This excited 4f state can radiatively emit: $4f^{n+1}L^{q-1} \rightarrow (4f^n)^*L^q \rightarrow 4f^nL^q + h\nu$.

Host-lattice excitation is a special type of excitation where an electron from the host is excited across the bandgap. The electron and hole can travel through the lattice in a correlated motion and can transfer their energy to the rare earth ion. This energy transfer leaves the rare earth ion in an excited state, which again can lead to emission from the rare earth.

Quantum yields are to a great degree determined by interaction with the host through the host's vibrational frequencies. The hosts most rare earths are doped in can have vibrational frequencies that are sufficiently low, less than 4-5 times the energy difference of the energy difference of the

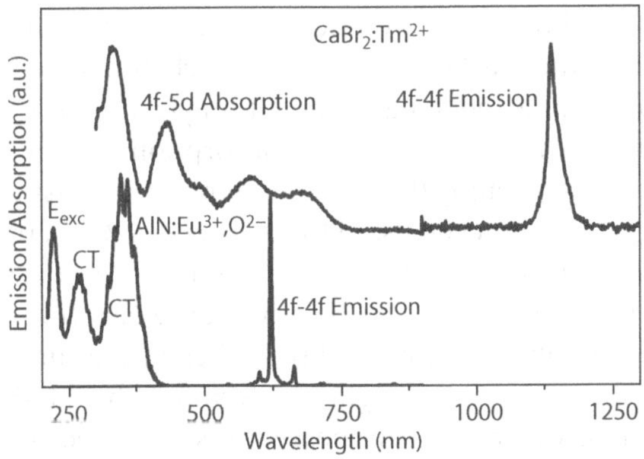

Figure 6.5 Optical transitions in $CaBr_2:Tm^{2+}$ (top) and $AlN:Eu^{3+},O^{2-}$ (bottom). $CaBr_2:Tm^{2+}$ displays a very broad 4f-5d absorption region, spanning the entire visible spectrum, followed by 4f-4f emission at 1140 nm. $AlN:Eu^{3+}$ O^{2-} shows two other types of absorption: Exciton (E_{exc}) and two charge-transfer (CT) transitions. These are followed by a series of 4f-4f emissions, centered around 622 nm.

radiative relaxation [16, Section 4.2], leading to that only very little energy is lost to lattice vibrations. With little energy lost to lattice vibrations, all the remaining energy can be used for the generation of luminescence, which establishes itself as a high quantum yield.

6.2.3 Luminescence in Quantum Dots

Colloidal quantum dots, often abbreviated as quantum dots (QDs) in the luminescence community, are nanocrystals (NCs) of semiconductor material so small in size that their optical properties are dictated by quantum confinement effects, rather than by their bulk material properties. Because of these effects, QDs are often likened to artificial atoms, since, to a very good degree, the effects observed in QDs can be described with the same equations as the hydrogen atom (see, e.g., Refs. [24, 25]). When concerning a QD made of a single material, changing the diameter of the QD directly influences its optical properties. By altering the diameter of the QD, the emission wavelength can be changed. A larger diameter QD leads to a smaller bandgap and therefore a more redshifted emission. Literature often features images of vials of QDs, where each vial contains a solution of QDs with a specific diameter, showing that the emission wavelength is tunable over the entire visible spectrum. A disadvantage of QDs composed of single materials, such as CdSe of PbS, are their negligible Stokes shifts, causing the QD LSC to suffer greatly from self-absorption losses. Furthermore, due to the size of a quantum dot (typically only a few nanometers in diameter) there is a large surface-to-volume ratio. This means that there are many surface atoms, leading to a large fraction of the QD having unsaturated bonds. Such unsaturated bonds can lead to channels for non-radiative recombination, thereby reducing the quantum yield of the QD [26].

Enabling larger energy shift between absorption and emission in QDs can be achieved by separating the absorption from the emission by fabricating core/shell heteronanocrystals. In these core/shell QDs, the shell (made from a different material than the core) serves as an sensitizer, transferring its absorbed energy to the core, as seen in Figure 6.6. The core functions as activator, eventually emitting the absorbed energy as luminescence [13, 27]. Here, the absorption wavelength is determined by the shell thickness, in much the same way as the luminescence is governed by the diameter of the core. Adding a shell to the emitting core of the quantum dot does solve the problem the core dots have of unsaturated bonds at the surface. However, depending on the core/shell configuration chosen, it does introduce a different set of problems. With the addition of a shell, the QD grows

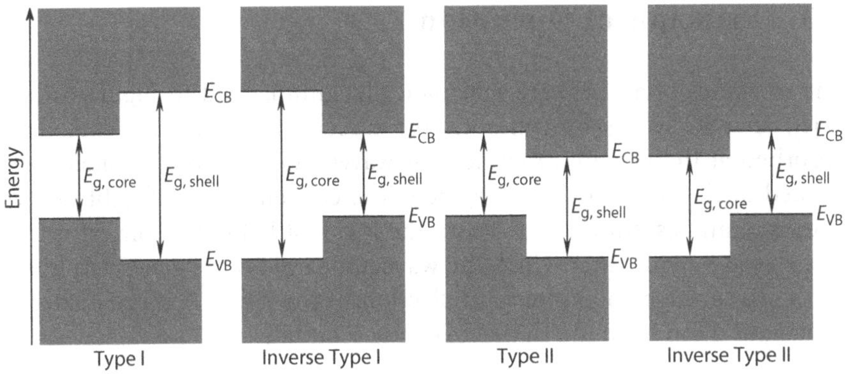

Figure 6.6 The different types of core/shell quantum dots [27].

in size, increasing the likelihood of light scattering and therefore a drop in efficiency. In the case of Type I core/shell QDs, the shell can still trap the charge carriers or have them leak to the surface, leading to reduced quantum yields [28]. In the case of Type II core/shell QDs, the hole and electron are spatially separated. While this does give the advantage of tuning the emission further to the infrared (avoiding self-absorption) than was possible with just the core or shell material, it does also mean that it will take some time for the spatially separated hole and electron to recombine radiatively. This increased radiative recombination time can make it possible for faster non-radiative quenching processes to dominate the luminescence, leading to lower quantum yields [26].

6.2.4 Hybrid Combinations

All luminescence processes described in the previous sections concern one family of luminophores. The processes involved all require transfer of energy from one state to another, which does not have to be confined to only one family of materials. Often combinations of two techniques are seen. For instance, rare earths can be combined with dyes to form an organometallic compound. In such compounds, the dye is responsible for absorption, but will transfer its energy to the rare earth, which is in turn responsible for the emission. Similarly, one can dope QDs with transition-metal [29] or rare earth ions [30]. Here, the semiconductor host absorbs energy, which it transfers to the doped cation. As the shift from absorption to emission is now governed by the dopant instead of the QD, a large, structure-independent shift is possible.

6.3 Principle of Operation

The workings of an LSC are schematically summarized in Figure 6.7. In brief, an LSC works by concentrating direct and diffuse incident light absorbed at the face of a luminescent waveguide onto photovoltaic cells located at the sides of the waveguide. The concentration of light occurs because luminescent centers absorb the incident light and convert it to a longer wavelength, after which the waveguide guides the generated luminescence through total internal reflection to the sides. A more in-depth explanation can be found in the following sections.

6.3.1 Absorption of Light

At its core, an LSC is nothing more than a luminescent material embedded in, or placed on top of a waveguide. For an LSC to function, it needs to be exposed to a source of light. An LSC works with both direct and diffuse light. Dependent on the refractive index of the material used to build up the LSC, light will be refracted into it. For an LSC made of a single slab of PMMA. 96% of light incident normal to the surface of the LSC will be refracted into the LSC. Light incident at different angles will have lower efficiencies in accordance with the Fresnel laws, as will be explained in Section 6.4.5. The light that has been refracted into the LSC can now be

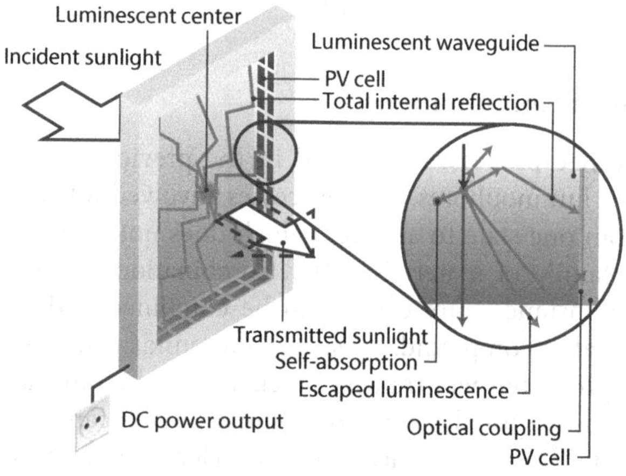

Figure 6.7 Schematic of the working of a single-layer LSC. Other configurations, such as (laminated) films on a waveguide (e.g., glass or polycarbonate), or tandem configurations, where two LSCs are stacked on top of each other, are also possible (Figure adapted from [31]).

absorbed by the luminescent particles within. Of course, not all refracted light will be absorbed. The amount of absorption is dependent on both (i) the physical properties of the luminescent particles, and (ii) design considerations of the LSC. For LSCs intended for use as a building-integrated photovoltaic window, an LSC intended to absorb ultraviolet (UV) light can feature up to 100% absorption, whereas LSCs that absorb visible light (up to a specific color, or the entire visible spectrum) should feature lower absorption, in order to still provide sufficient light for the environment behind the window. In a study by Vossen, Aarts, and Debije [32], it was shown that 25% absorption is the limit for red-tinted LSC BIPV windows. All things remaining equal, an LSC that absorbs 25% of the visible spectrum will always outperform an LSC that absorbs 100% of the UV, only due to sunlight having substantially more photons in the visible spectrum than in the UV. All light that was refracted into the LSC and not absorbed will be refracted out the back of the LSC and illuminate the area behind it.

6.3.2 Emission within the LSC

The light absorbed by the luminescent particles will be converted to emission characteristic for the luminescent particle in question. Usually, especially with experimental materials, this conversion does not happen with 100% efficiency. Part of the absorbed light will excite the available phonon modes in the LSC, eventually only generating heat. The absorbed light that "survived" conversion will now be emitted isotropically within the LSC.[7] This separation of absorption and emission is precisely what makes an LSC independent of the direction and type of incident light, as the direction of the emitted photons is randomized independently of their origin. As happened with the light incident of the LSC, the light emitted by the luminescent particles can be refracted out of the LSC's faces. Again, this happens in accordance with the Fresnel laws. The light that is refracted out is said to "be in the escape cone." The light that is not refracted out of the LSC will remain "trapped" within the waveguide by total internal reflection (TIR) and will start traveling through the LSC, much like light coupled into an optical fiber would. The percentage of converted photons trapped within the LSC is referred to as the trapping efficiency η_{trap}. For a single-slab PMMA LSC, η_{trap} amounts to approximately 75%.

---

[7]As with everything, exceptions to this generalization exist. A subclass of LSC materials feature non-isotropic emission with, for example, specially oriented nanorods that are optimized to couple the emitted light to the sides of the LSC, promoting total internal reflection, and therefore overall efficiency [33, 34].

6.3.3 Effects of Self-Absorption

Depending on the LSC material used, what is described in the previous paragraphs would be most of what happens within the LSC itself. However, if the used luminophore has an overlap between its absorption and emission spectrum, the emitted light traveling through the LSC can be reabsorbed in a process known as parasitic self-absorption. As can be seen for BASF Lumogen F Red305 in Figure 6.3, one of the best-known LSC materials, quite a degree of self-absorption can occur. Parasitic self-absorption is a problem, as it results in a photon that was traveling to the LSC perimeter by TIR becoming reabsorbed by the luminescent particles. The photon then might get quenched (non-unity η_{QY}) or get re-emitted, but in such a direction that it is lost in the escape cone. It should, however, be noted that self-absorption does not always have to be detrimental to the overall efficiency of an LSC. First shown theoretically by Merkx, Kate, and Kolk [35] and then experimentally by Krumer *et al.* [36], if a material has a sufficiently high η_{QY}, paired with a sufficiently high absorption, self-absorption can bolster the overall LSC efficiency, as shown in Figure 6.8. If the absorption is sufficiently high, light originally emitted in the escape cone will be reabsorbed and subsequently re-emitted with a high probability of not ending up in the escape cone again, thereby contributing to the overall LSC efficiency.

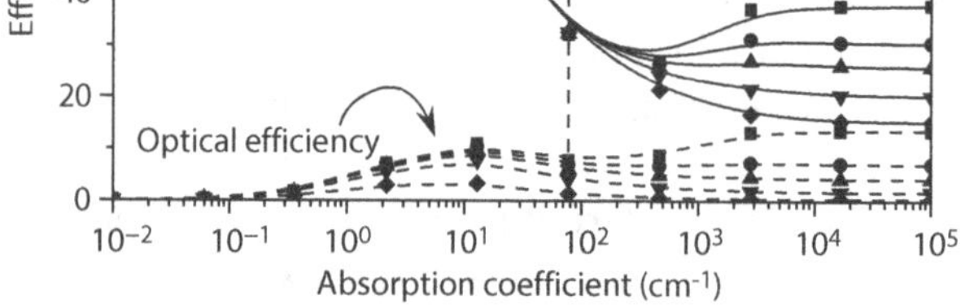

Figure 6.8 Theoretical efficiencies of a Red305 LSC with varying absorption coefficient and quantum yield. At a sufficiency high quantum yield, the trapping of light can increase due to light being recaptured in the escape cone. This recaptured light can then partake in the total internal reflection process [35]. The optical efficiency rises, since with an increasing absorption coefficient the absorption of incident sunlight rises.

6.3.4 Influence of the Waveguide

Even with luminophores that have no overlapping absorption and emission, another constraint on the light transport efficiency is the waveguide itself [37, 38]. Scattering of light during TIR is a loss mechanism in LSCs. This scattering can happen either at the surface of the waveguide due to surface roughness, or within the waveguide due to defects within the LSC itself. This scattering leads to a randomization of the traveling direction of the formerly waveguided light and therefore the light might be scattered into the escape cone.[8] For LSCs > 1 m^2, such as LSC-based windows, the eventual device efficiency can differ by a factor 10, based on the choice of waveguide. As shown in Ref. [13], normal window glass (soda-lime) has an absorption that is too high in the red to near infrared spectrum relevant to LSCs to be of practical use in windows. Other materials, such as N-BK7 glass, fluorinated PMMA, or TIREXtreme, make a much more viable non-absorbing waveguide. For IPV applications, the absorption from the waveguide will not impact performance much, given the short distances emitted light has to travel.

6.3.5 Conversion of Concentrated Light to Electricity

When the TIR light reaches the perimeter of the LSC, most conventional LSC designs dictate that the light will be coupled out of the waveguide, mediated by a form of optical coupling, to finally be converted into electricity by a photovoltaic cell. Most often, off-the-shelf Si PV cells are used in LSC research [40], since most contemporary LSCs output yellow to red light, where a Si PV cells has excellent performance. However, off-the-shelf photovoltaics are to an increasing degree optimized to absorb the solar spectrum, just the same way LSCs are. LSCs try to convert this sunlight outside of their absorption range, i.e., out of the visible range and into the red to NIR, where Si PV cells start to lose performance.

One strategy to overcome the spectral mismatch between LSC output and Si input is to use high efficiency III-V PV cells, such as GaAs [10] or InGaP [41, 42], which show that the LSC efficiency can be greatly enhanced by using better matched cells (typically by a factor 2–3 [35]). Another approach would be to latch on to the recent developments in wavelength-tunable PV, such as perovskite-based solar cells, or by tuning

[8] Here it is also the case that sufficient scattering might lead to an increase in LSC performance [39]. A high degree of scattering also means that the LSC will appear hazy, defeating its use as a window-like PV.

the bandgap of CGS cells that have better absorption characteristics in the LSC output region [43].[9]

6.4 Calculating LSC Performance

With the qualitative description of the previous section in mind, an LSC can be modeled in much the same way as one would model a PV module: through ray tracing. As such, some LSC articles even add to conventional PV modeling software or principles. This ray tracing approach will be treated independently of software packages in Section 6.4.5. More analytic or semi-analytic approaches that follow the same paradigm of tracing the separate optical steps are described for "back-of-the-envelope" calculations that yield the maximal possible output for an LSC with a simple geometry are described in Section 6.4.2 and Section 6.4.3. A more detailed approach suitable for any LSC geometry is described using a quick semi-analytic approach in Section 6.4.4. In addition to the methods presented here, LSCs can also be described through entropic models, as described extensively in Ref. [45].

6.4.1 Figures of Merit

Before we start with how an LSC can be calculated in full, we first have to define what it exactly is that we want to calculate. The goal of an LSC is to convert light incident on its faces to light radiating from its sides. This concentration of light is quantified by what in literature is referred to as the optical efficiency

$$\eta_{\text{opt}} = \frac{\#\,\text{photons in}}{\#\,\text{photons out at edges}} = \frac{\int S_{\text{in}}(\lambda)\,d\lambda}{\int S_{\text{edges}}(\lambda)\,d\lambda}. \tag{6.2}$$

η_{opt} is sometimes also referred to as the LSC luminescence efficiency $\eta^{\text{PL,LSC}}$ [46], and sometimes, confusingly, the external quantum efficiency [41], although the external quantum efficiency can also refer to all light emitted from the LSC (i.e., both from the faces and the sides).

---

[9]A more detailed discussion on the merit of solar cells other than Si for use as IPV can be found in Chapter 3 or in Ref. [44].

Here it should be noted that η_{opt} is not concerned with the origin of the photons or whether their wavelengths are useful for further conversion by a photovoltaic cell. This is where the PV, or power efficiency, comes in:

$$\eta_{\text{Power}} = \frac{I_{SC} V_{oc} FF}{P_{in}}. \tag{6.3}$$

Here, I_{sc} is the short-circuit current, V_{oc} the open-circuit voltage and FF the filling factor of the LSC waveguide combined with photovoltaic cells while exposed to a light source (such as the sun, an LED, or tube lighting) with power P_{in}.

In LSC literature, the optical efficiency is preferred over the power efficiency. The power efficiency is too dependent on the performance of the optical coupling and the types of PV cell used and therefore does not say much about the performance of the LSC itself.

To quantify if using an LSC is more efficient than directly exposing the attached PV cells to the light source, another figure of merit is introduced: the concentration factor

$$\Gamma = \frac{G\eta_{\text{Power}}}{\eta_{\text{PV}}}, \tag{6.4}$$

where $G = \frac{A_{LSC}}{A_{PV}}$ is the geometric gain (the ratio of the face area of the LSC to that of the attached PV cells) and η_{PV} the efficiency of the used PV cells.

At the time of writing, the top-performing single-crystal Si PV cells have $\eta_{PV} = 26.7\%$ [47, 48].

If we assume a window measuring $1 \times 1 \times 0.05\ \text{m}^3$, this would mean that for an LSC with PV cells all around the window frame, $A_{PV} = 0.2\ \text{m}^2$. The LSC itself covers the entire window, measuring at $A_{LSC} = 1\ \text{m}^2$. For an LSC to make sense, $\Gamma \gg 1$. Therefore, $\eta_{\text{Power}} \gg 5.3\%$ for it to make sense to install an LSC, as opposed to exposing the PV cells directly to the incident sunlight.

6.4.2 Upper Bound for LSC Efficiency

In its most essential form, the behavior of an LSC can be summarized as

$$\eta_{opt} = (1-R)\, \eta_{abs}\, \eta_{QY}\, \eta_{trap}\, \eta_{SA}\, \eta_{WG}. \tag{6.5}$$

Here, R is the amount of reflection from the LSC face, η_{abs} is the fraction of light that the luminophores contained within the LSC are able to absorb, η_{QY} is the quantum yield of the used luminophores, η_{trap} is the fraction of light emitted outside the escape cone and η_{SA} and η_{WG} are factors to take respectively the self-absorption and waveguiding (scattering) efficiency into account.

An upper bound for this optical efficiency, which does not take self-absorption into account, is given by

$$\eta_{opt} = \frac{\int [1-R(\lambda)]\overbrace{S(\lambda)[1-\exp(-\varepsilon(\lambda)cd)]}^{\eta_{abs}}\eta_{QY}\overbrace{\sqrt{1-\frac{1}{n_{LSC}^{2}}}}^{\eta_{trap}}\,d\lambda}{\int S(\lambda)\,d\lambda} \tag{6.6}$$

Here, the first braced term is the Beer-Lambert law as a function of the molar attenuation coefficient $\varepsilon(\lambda)$ of the luminophore at concentration c for an LSC of thickness d for a light source with spectrum (expressed in amount of photons) $S(\lambda)$, usually taken to be the AM1.5g spectrum, when discussing LSCs for solar (outdoor) applications. The term in the second brace is η_{trap} written out according to Snell's law for a refractive index of η_{LSC}, for an LSC surrounded by air at both faces.

We now have all the ingredients needed to see what a hypothetical LSC should be capable of in order to be more effective than directly exposing the perimeter's PV cells to sunlight. As we saw in Section 6.4.1, for a $1 \times 1 \times 0.05\ m^3$ LSC with a perimeter of Si solar cells, we want $\eta_{Power} \gg 5.3\%$. To accomplish this, we can define two types of LSCs that would not cause artificial color blindness, shown in Figure 6.9a. One type absorbs uniformly until 400 nm: a UV-absorbing LSC that leaves the visible spectrum unaltered. The other absorbs uniformly until 780 nm, thereby also dimming the visible light that enters a building. Both LSCs feature no self-absorption (as we cannot account for that yet) by emitting a 20 nm full-width half-maximum Gaussian centered at 900 nm. These hypothetical LSCs furthermore have $n = 1.49$ (PMMA), yielding $(1 - R) = 0.96$ for light at normal incidence and $\eta_{trap} = 0.75$.

Calculating the power efficiency for such an LSC goes in much the same way as one would calculate the theoretical efficiency for a solar cell. The theoretical power efficiency for a solar cell is found by integrating the external quantum efficiency $EQE(\lambda)$ of the PV cell multiplied by the spectrum of light incident on the PV cell. In our case, that is the edge-emission spectrum of our LSC $\Phi_{em}(\lambda)$ in terms of photons per second. All these

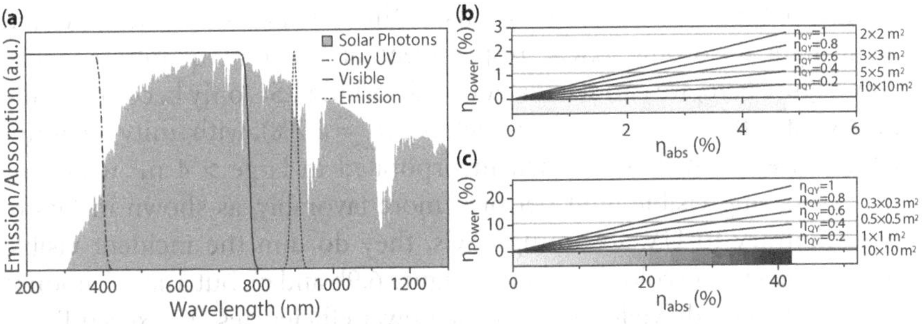

Figure 6.9 Power efficiencies for two hypothetical LSCs without self-absorption, (a) Absorption spectra for two LSCs, one absorbing only UV light (until 400nm), one absorbing UV and all visible light (until 780nm), both emit at 900 nm. Behind are normalized photon counts for AM1.5g. (b), (c) Power efficiencies for the UV (panel b) and visible (panel c) absorbing LSC. Grey reference lines are placed indicating at what the surface area these square LSCs would exceed T = 1. Below the graphs is an indication of the perceived brightness of light passing through at the specified absorption efficiency.

photons will get an energy equal to the bandgap of the PV cell E_g, which for Si is 1.12 eV. Excess energy of photons will be converted to heat.

The power efficiency is ratio of the power output yielded by the converted photons to the incident solar power (in this case with an AM1.5g spectrum $S_{AMI.5}$ (λ), expressed in photon flux). This leads to

$$\eta_{\text{Power}} = \frac{\int_{\lambda=0}^{\infty} EQE(\lambda) \times E_g \times \Phi_{\text{em}}(\lambda)\mathrm{d}\lambda}{A_{\text{LSC}} \int_{\lambda=0}^{\infty} S_{\text{AM1.5}g}(\lambda)\left(\dfrac{hc}{\lambda}\right)\mathrm{d}\lambda}, \tag{6.7}$$

with $EQE(\lambda)$ the external quantum efficiency and E_g the wavelength-dependent external quantum efficiency and the bandgap of the used PV cell, respectively. $\Phi_{em}(\lambda)$ and A_{LSC} are the edge-emission spectrum in terms of photons per second and the area of the LSC, respectively, $\frac{hc}{\lambda}$ is the energy-conversion factor (1240/λ eV) from photon counts to energy.

Without self-absorption, the flux of converted photons Φ_{em} can be easily retrieved from the optical efficiency, combined with the emission spectrum of the LSC S_{PL} (λ):

$$\Phi_{\text{em}}(\lambda) = \underbrace{\eta_{\text{opt}} A_{\text{LSC}} \int_{\lambda=0}^{\infty} S_{\text{AM1.5g}}(\lambda)\mathrm{d}\lambda}_{\text{Rate of photon conversion}} \underbrace{\frac{S_{\text{PL}}(\lambda)}{\int_{\lambda=0}^{\infty} S_{\text{PL}}(\lambda)\mathrm{d}\lambda}}_{\text{Normalized emission}} \tag{6.8}$$

Figure 6.9b and c show the results of this calculation for our UV and visible light absorbing LSCs, respectively, at different quantum yields. From Figure 6.9b, we can see that a UV absorbing LSC only becomes really effective at 100% UV-absorption (yielding η_{abs}= 4.7%), with unity quantum yield; and even then only when incorporated in large > 4 m^2 windows. LSCs absorbing visible light perform more favorably, as shown in Figure 6.9c. Contrary to UV-absorbing LSCs, they do dim the incident visible light (as seen in the color bars below Figure 6.9b and c), but a small amount of dimming already yields appreciable power efficiencies. To exceed $\Gamma = 1$, for a 1 × 1 × 0.05 m^3 LSC with a perimeter of Si solar cells at η_{QY}= 1, an η_{abs} of 12% is needed, which amounts to only a minor amount of dimming, as can be seen in Figure 6.9c. Increasing the window size quickly yields $\Gamma > 1$ for LSCs absorbing the entire visible spectrum.

6.4.3 Analytical Approach for Simple Geometries

When considering self-absorption losses, the geometric-progression model developed by Klimov *et al.* can be used when dealing with an LSC with collection at two edges of length *L*, and mirrors at the other two edges [49]. Klimov *et al.* took for this model the analytical equations first introduced by Weber and Lambe [1] in their seminal paper. They simplified the model of Weber and Lambe, which does not take reemission of self-absorbed light into account, by fitting the model's self-absorption efficiency $\eta_{SA,W\&L}$ with

$$\eta_{SA,W\&L} = \frac{1}{1+\beta\alpha_2 L} \tag{6.9}$$

which resulted in $\beta = 1.4 + 0.5\sqrt{L/150}$ from fitting with

$$\alpha_2 = \frac{\int S_{PL}(\lambda)\alpha(\lambda)\,d\lambda}{\int S_{PL}(\lambda)\,d\lambda} \tag{6.10}$$

the average absorption coefficient of emitted light $S_{PL}(\lambda)$ by an LSC with Napierian absorption coefficient $\alpha(\lambda)$[10].

To account for self-absorption, the model uses the notion that the amount of light concentrated at the LSC perimeter is the light resulting

[10]In the case of sharp absorption edges, this equation will yield unexpected results. In such a case, a non-averaged α2 should be used instead.

from conversion by the luminophores, except for the light that is lost due to reabsorption:

$$\eta_{\text{con}}^{(0)} = \eta_{\text{SA,W\&L}} \eta_{\text{QY}} \eta_{\text{trap}}. \tag{6.11}$$

Light that is reabsorbed is converted again, with an efficiency

$$\eta_{\text{con}}^{(1)} = \eta_{\text{QY}} \eta_{\text{trap}} (1 - \eta_{\text{SA,W\&L}}) \eta_{\text{QY}} \eta_{\text{con}}^{(0)}. \tag{6.12}$$

Extending this for every reabsorption event leads to

$$\eta_{\text{con}}^{(i)} = (\eta_{\text{QY}} \eta_{\text{trap}})^{i+1} (1 - \eta_{\text{SA,1}})^{i} \eta_{\text{SA,W\&L}} \tag{6.13}$$

and therefore, filling in the expression for $\eta_{\text{SA,W\&L}}$ from Eq. (6.9):

$$\eta_{\text{con}} = \sum_{i=0}^{\infty} \eta_{\text{con}}^{(i)} = \frac{\eta_{\text{QY}} \eta_{\text{trap}}}{1 + \beta \alpha_2 L (1 - \eta_{\text{QY}} \eta_{\text{trap}})} \tag{6.14}$$

with, as mentioned before, $\beta = 1.4 + 0.5\sqrt{L/150}$.

This leads to a simple expression for the total optical efficiency:

$$\eta_{\text{opt}} = \frac{\int [1 - R(\lambda)] S_{\text{in}}(\lambda) \{1 - \exp[1 - \alpha(\lambda) d]\}\, d\lambda}{\int S_{\text{in}}(\lambda) d\lambda} \eta_{\text{con}}$$

$$= \frac{\int [1 - R(\lambda)] S_{\text{in}}(\lambda) \{1 - \exp[1 - \alpha(\lambda) d]\}\, d\lambda}{\int S_{\text{in}}(\lambda)\, d\lambda} \frac{\eta_{\text{QY}} \eta_{\text{trap}}}{1 + \beta \alpha_2 L (1 - \eta_{\text{QY}} \eta_{\text{trap}})} \tag{6.15}$$

A disadvantage of this model is its limited use cases, being only valid for an LSC with two mirroring faces.

6.4.4 Semi-Analytical Optimization Calculations for Arbitrary Geometries

Extending the notion of including self-absorption, a semi-analytic model to calculate LSC performance for cylindrical LSCs was introduced by

Kate *et al.* [50]. This model was later extended by Merkx *et al.* [35] for all geometries. This model takes self-absorption into account by modeling the LSC with several "generations" of light absorption steps, hence the name of the model: Multiple Generation Light Transport (MGLT). What makes this approach different from the geometric-progression model presented by Klimov *et al.* is the possibility to calculate every aspect of LSC light transport, not limited to just the device efficiency, for LSCs of arbitrary geometries. Parameters such as escape cone losses, the exact shape (in terms of intensity per wavelength) of the edge-emission spectrum, and the degree of self-absorption of the LSC as a function of position can be calculated. In the following, a summarized description of the used equations is presented. After that follows a short example of how the model works for a dye-based LSC with strong self-absorption. To further get a feel for how the model works and how the equations involved influence the LSC efficiencies, a worked out example in MATLAB code of the MGLT model is available in Ref. [51]. For this model, only the absorption and emission spectra, the quantum yield, and the refractive index of the LSC of interest are required as input parameters.

In the MGLT model, the optical efficiency is calculated over a series of light emission generations N_G

$$\eta_{\mathrm{opt}} = \frac{(1-R)\eta_{\mathrm{abs}}}{N_{\mathrm{absorbed}}} \sum_{i=0}^{N_G} \eta_{\mathrm{trap}} \lambda \iint_{A_{\mathrm{LCS}}} \sigma_{\mathrm{i}} - \frac{1}{\eta_{\mathrm{trap}} + \eta_{\mathrm{CA}}} \gamma_{\mathrm{i}} \, dxdy. \quad (6.16)$$

Here, $N_{\mathrm{absorbed}} = (1 - R)$, $\eta_{\mathrm{abs}} N$ are the amount of absorbed photons from a total of N incoming photons, σ_i is the emission of the luminophores of generation i, while $\frac{1}{\eta_{\mathrm{trap}} + \eta_{\mathrm{CA}}} \gamma_i$ is the absorption of that generation, including any photons in the escape cone that are reabsorbed before they can exit the waveguide.

For every generation i the difference between emission and absorption is calculated taking escaped or quenched photons into account, similar to the geometric series model as presented in Section 6.4.3. The main difference between these models is how exactly the absorption is calculated. In the MGLT model, absorption γ_{i} of a generation i is given by the two-dimensional convolution of the absorption density $a(x, y, \lambda)$ and the emission of that generation $\sigma_i(x, y, \lambda)$:

$$\gamma_i (x, y, \lambda) = [\eta_{\mathrm{trap}} + \eta_{\mathrm{CA}} (\lambda)] \sigma_i (x, y, \lambda) ** a (x, y, \lambda), \quad (6.17)$$

with

$$a(x,y,\lambda)=\frac{\alpha'(\lambda)\exp\left[\alpha'(\lambda)\sqrt{x^2+y^2}\right]}{2\pi\sqrt{x^2+y^2}}, \tag{6.18}$$

where $\alpha'(\lambda)$ is an "effective" absorption. This effective absorption corrects the absorption spectrum in such a way that the path that light travels through the LSC is independent of the polar emission angle θ:

$$\alpha'(\lambda)=\frac{\alpha(\lambda)}{\pi-2\theta_c}\int_{\theta_c}^{\pi-\theta_c}\frac{1}{\sin(\theta)}\,d\theta. \tag{6.19}$$

Eq. (6.17) corrects for absorption in the escape cone with

$$\eta_{CA}=\frac{1}{d}\int_0^d\int_0^{\theta_c}1-\exp\left[-\frac{\alpha(\lambda)z}{\cos(\theta)}\right]\sin(\theta)\,d\theta dz. \tag{6.20}$$

Finally, the (re)emission spectra of the LSC are given by

$$\sigma_i(x,y,\lambda)=\begin{cases}\dfrac{\eta_{QY}S_{PL}(\lambda)}{\lambda}N_{absorbed}, & \text{for } i=0.\\[2ex] \dfrac{\eta_{QY}S_{PL}(\lambda)}{\lambda}\displaystyle\int_{\lambda=0}^{\infty}\lambda\gamma_{i-1}(x,y,\lambda)d\lambda, & \text{for } i>0.\end{cases} \tag{6.21}$$

A good way to understand what is happening in the equations above is to do an example calculation. For this example calculation we will use the dye Lumogen Orange (94738 Fluoreszenzorange) [52] embedded in a sheet of $10 \times 10 \times 0.5$ cm^3 PMMA ($n = 1.49$), with absorption and emission spectra as shown in Figure 6.10a. This dye has a very high quantum yield $\eta_{QY} = 0.95$ and features strong self-absorption.

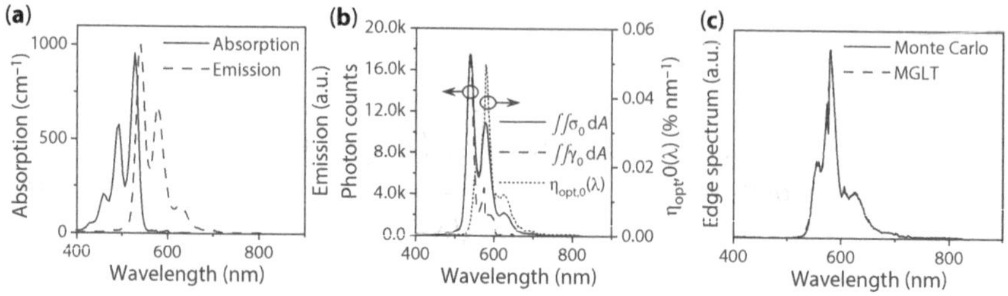

Figure 6.10 Properties of Lumogen Orange [52]. (a) Absorption and emission spectra, (b) Results for the first generation of MGLT. (c) Light spectrum observed at the edges of the LSC, calculated with both MGLT and Monte Carlo. The lines overlap perfectly, without any scaling.

We start at generation $i = 0$. To calculate the optical efficiency, we need to see what is absorbed. For our LSC, this amounts to

$$\frac{N_{\text{absorbed}}}{N} = \frac{\int [1-R(\lambda)]\, S(\lambda)\{1-\exp[1-\alpha(\lambda)d]\}\, d\lambda}{\int S(\lambda)\, d\lambda} = 6.88\% \qquad (6.22)$$

This absorbed light will then be reemitted by the luminophore in accordance with Eq. (6.21). Looking at Eq. (6.16), the only thing still required to calculate the optical efficiency of this first generation of light transport is the amount of reabsorption of this emitted light, given by Eq. (6.17). The results for emission and reabsorption in the first generation integrated over the surface of our LSC are plotted for the input spectrum from Figure 6.10a in Figure 6.10b. From the difference in emission and reabsorption, we can calculate the edge spectrum $\eta_{\text{opt},0}$ (λ) for the first generation. For the first generation, this is shown on the right axis of Figure 6.10b, with a more detailed image for all generations visible in Figure 6.10c. Integrating $\eta_{\text{opt},0}$ (λ) over all wavelengths, gives us the solution to Eq. (6.16) with $N_G = 1$. This integration yields a total optical efficiency for the first generation of light transport of 2.14%. If we increase the number of generations, each generation will progressively contribute less, as seen in Table 6.1, until culminating in a total optical efficiency of $\eta_{\text{opt}} = 3.76\%$.

A practical usage of the MGLT model is for the precise optimization of planar LSCs, using only a limited number of light transport generations. For optimization, this model is preferable to Monte Carlo simulations (detailed in the next section), because of the constant time it takes to calculate each generation, regardless of LSC geometry or absorption strength. Furthermore, this model has been shown to agree with Monte Carlo

Table 6.1 Contribution to optical efficiency per generation for a Lumogen Orange LSC plate simulated with MGLT. The simulation was stopped after simulating 12 generations, with the total contributions of generation 8 to 12 adding to less than 0.01%.

Generation i	Contribution to optical efficiency (%)
0	2.14
1	0.92
2	0.40
3	0.17
4	0.08
5	0.03
6	0.01
7	0.01
8… 12	0.00
Total	3.76

simulations to an excellent degree, as shown in Figure 6.10c. This agreement serves as a kind of "litmus test" for the validity of the MGLT model. Not only is the eventual calculated optical efficiency (η_{opt} = 3.76%) identical for both simulations, but even the details of the edge emission spectrum.

6.4.5 Monte Carlo Simulations for Ray-Traced Complex Geometries

In addition to all (semi)analytical models presented in the previous sections, there is the industry standard way of simulating LSC behavior: Monte Carlo (MC) simulations. In MC simulations, individual incident photons are simulated, but the wave nature of light is ignored in favor of computationally more efficient and highly parallelizable ray tracing. A great advantage of using MC opposed to analytical models is that any geometry LSC can be taken into account, as well as involved waveguide designs, without having to alter any equations, contrary to what would be the case with the previously described analytical models. A disadvantage of MC when compared to analytical models is that MC simulations do not run in constant time, and may take a long time to converge to accurate results [53, Ch. 3]. Several commercial (e.g., Light Tools [54] or COMSOL [55]) and non-commercial (e.g., PVTrace [56] or MC.m [35, 57] for unpolarized light) MC ray tracing software products exist, making it easy to set up and simulate an LSC.

MC simulations approach the problem of ray tracing stochastically. That means that an individual ray can only fully transmit, reflect, get absorbed, etc., and that only when looking at the average of all simulated rays something can be said about, for instance, partial absorption of incoming light.

An MC simulation is initialized by generating a photon incident on the LSC surface. This photon has a polarization (s or p), azimuthal angle φ, polar angle θ and position in space $\vec{x}$. For simulations of LSCs, it is efficient to have the initial position to be somewhere on the face of the LSC; for a rectangular LSC of dimensions $l \times w \times d$:

$$\vec{x} = \begin{pmatrix} x \\ y \\ z \end{pmatrix} = \begin{pmatrix} \xi_x \omega \\ \xi_y l \\ d \end{pmatrix}, \tag{6.23}$$

with ξ a random number between 0 and 1, either random or chosen from a Halton sequence [53, Ch. 3]. For most simulations, light incident on the LSC is chosen to be normal to the LSC surface, i.e., $\varphi = 0$, $\theta = \pi$ [58, 59].

A photon is reflected or transmitted at the LSC interfaces in accordance with Snell's law

$$n_1 \sin(\theta_1) = n_2 \sin(\theta_2) \tag{6.24}$$

For transmission, these will be combined with the Fresnel equations

$$\tau_s = \frac{2n_1 \cos(\theta_1)}{n_1 \cos(\theta_1) + n_2 \cos(\theta_2)} (\text{s}-\text{polarization}) \tag{6.25}$$

$$\tau_p = \frac{2n_1 \cos(\theta_1)}{n_1 \cos(\theta_2) + n_2 \cos(\theta_1)} (\text{p}-\text{polarization}) \tag{6.26}$$

leading to a probability of transmission for either polarization

$$T = \frac{n_2 \cos(\theta_2)}{n_1 \cos(\theta_1)} |\tau|^2. \tag{6.27}$$

Similarly, for reflection

$$\rho_s = \frac{n_1\cos(\theta_1) - n_2\cos(\theta_2)}{n_1\cos(\theta_1) + n_2\cos(\theta_2)}(\text{s}-\text{polarization}) \tag{6.28}$$

$$\rho_\text{p} = \frac{n_1\cos(\theta_2) - n_2\cos(\theta_1)}{n_1\cos(\theta_2) + n_2\cos(\theta_1)}(\text{p}-\text{polarization}) \tag{6.29}$$

with probability of reflection for either polarization

$$R = |\rho|^2. \tag{6.30}$$

The most common occurrence for LSCs is exposure to sunlight. Sunlight can be considered to be unpolarized. This greatly simplifies the above equations, as the contributions of s and p polarization can simply be averaged:

$$R = \frac{R_s + R_p}{2} \cdot (\text{unpolarized light}) \tag{6.31}$$

In MC ray tracing, to see if a ray is reflected, a random number $\xi_R \in [0,1)$ is generated. If $\xi_R < R$, the light ray is reflected. Otherwise, the ray is transmitted to the next medium. The MC algorithm is explained schematically in Figure 6.11.

If the photon has entered the waveguide, the maximal path length ℓ, it can travel before absorption is given by

$$\ell = -\frac{1}{\alpha(\lambda)}\log(1-\xi_\ell), \tag{6.32}$$

with $\xi_\ell \in [0,1)$.

Combined with the Fresnel equations, the new position of the photon can be calculated. If the photon is absorbed within the LSC, we have to check if luminescence occurs. This is the case if $\xi_{QY} < \eta_{QY}$, with $\xi_{QY} \in [0,1)$. In the event of luminescence, the photon will be assigned a new wavelength

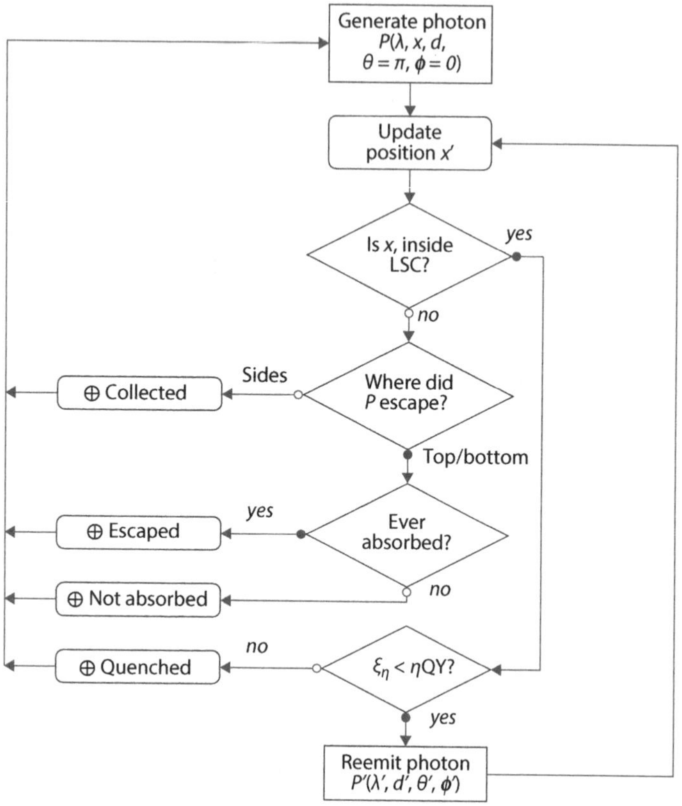

Figure 6.11 A general schematic representation of the MC ray tracing flow. Reprinted from [35].

λ' by inverse transform sampling the emission spectrum of the LSC. Here, the emission spectrum is assumed a probability density function for the wavelength our photon will have upon emission. The angles under which the photon is reemitted are obtained using

$$\phi' = 2\pi\xi_{\phi}, \tag{6.33}$$

$$\theta' = \cos^{-1}(1-2\xi_{\theta}). \tag{6.34}$$

Here, $\xi_{\phi}, \xi_{\theta} \in [0,1)$.

Following reemission, we have to check again where the photon will head, solve the Fresnel equations upon interaction with an interface, or if the photon is absorbed on its path, check if reemission happens, until the photon has either escaped the waveguide or is quenched.

For unpolarized light traveling in an LSC consisting of a single slab of material, the transport calculation can be sped up. In this case, the new position $\vec{x}\lambda$ of our photon after traveling a distance ℓ at angles θ, ϕ can be found with

$$\vec{x}' = \vec{x} + \begin{pmatrix} \dfrac{\ell \sin(\phi)}{\sin(\theta)} \\ \dfrac{\ell \cos(\phi)}{\sin(\theta)} \\ \Delta z \end{pmatrix}. \tag{6.35}$$

Here, Δz is dependent on whether total internal reflection happens. In the case of total internal reflection ($\theta_c < \theta' < \pi - \theta_c$), Δz depends on the total number of reflections

$$N_{refl} = \frac{|\ell \cos(\theta) + z|}{d}. \tag{6.36}$$

Which yields for Δz:

$$\Delta z = \begin{cases} |\ell \cos(\theta') + z| \bmod (d) - z, & (\theta - \frac{\pi}{2}) \times Ev(N_{refl}) > 0 \\ d - |\ell \cos(\theta') + z| \bmod (d) - z, & Otherwise \end{cases} \tag{6.37}$$

where $Ev(N_{refl})$ returns 1 if N_{refl} is even, and -1 otherwise. In the case emission happens inside of the escape cone (no total internal reflection):

$$\Delta z = \ell \cos(\theta). \tag{6.38}$$

When the photon exits the LSC through the sides, it is assumed to be collected by the PV cells. Otherwise, the photon has escaped through the escape cone or has never been absorbed at all.

These ray tracing simulations can give direct insight into how light rays propagate through an LSC. Also, the amount of self-absorption can be

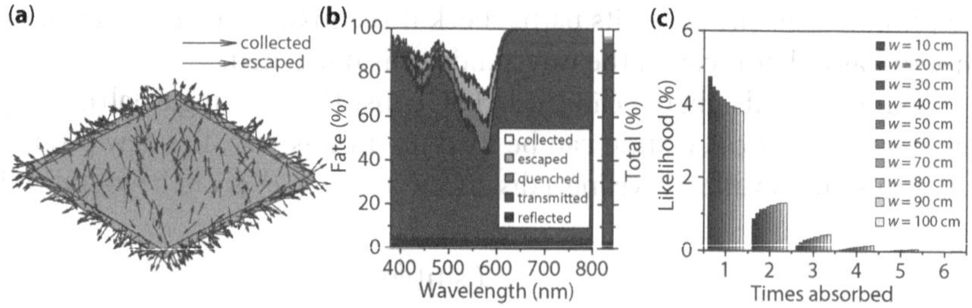

Figure 6.12 MC simulation for a hypothetical Red305 LSC with 80% visible transmission and $\eta_{QY} = 0.8$. (a) Simulation output giving an indication of where in the LSC light escapes and where it is collected. Here only a limited number of rays are displayed for clarity, (b) Distribution of where photons incident on the LSC end up. The wavelengths are the wavelengths of the incident light, (c) Number of reabsorption steps a photon undergoes in LSCs of differing sizes ($(l = w) \times w \times 0.5$ cm^3). 0 times absorbed is not displayed, as that value is determined by the transmission, and therefore the same for all LSC sizes (80%). For LSCs of increasing sizes, the number of absorption events scales with the size.

quantified by looking at the number of absorption events a ray is likely to undergo. This effect can be seen for a hypothetical Red305 LSC with 80% transmission of visible light and a quantum yield of 80% in Figure 6.12.

6.4.6 Considerations for Thin-Film LSCs

All presented models and equations in the previous sections neglected the wave nature of light. This assumption is reasonable, since all dimensions in PMMA LSCs (> 1 mm) and waveguides laminated with a luminescent coating (> 100 μm) exceed the wavelengths of sunlight under consideration (~ 300 nm to 1200 nm). In thin films (0.1 μm to 10 μm) this assumption does not hold any longer, and the wave nature of light must be considered. For thin films, the absorption, transmission, and reflection of light are altered by constructive and destructive interference effects. These interference effects can be taken into account when modeling LSCs using the transfer matrix method (TMM). Many software packages are available that handle the equations of a thin layer stack. The work by Byrnes [60] gives a good overview of what mechanics are at play when employing TMM for transmission, absorption, and emission calculations. For LSC work, one can assume that the luminophores are randomly distributed throughout the thin film LSC. This random distribution means that light transport can be assumed to happen incoherently through the waveguide [61]. Therefore, after calculating the amount of absorption using TMM, the methods presented in the previous sections can again be employed to calculate LSC efficiencies.

6.5 State-of-the-Art LSC Materials

6.5.1 Measures for the Visual Performance of LSC Materials

As is evident from the previous sections, an ideal LSC is the result of a careful balancing exercise. To bolster efficiency, as much light as possible should be absorbed, but this absorption should have no overlap with the emission, as self-absorption is mostly detrimental to performance. For windows, high absorbance in the visible spectrum should be avoided because of their application as light sources: an opaque plate does not make for a good window. Next to the desired low absorption in the visible spectrum, the absorption spectrum should be featureless, so the perceived color of objects illuminated through the window is still close to what it would be were the object illuminated with white light.

A measure for the perceived color is the color rendering index R_a, introduced by the International Commission on Illumination [62]. R_a compares the spectrum measured from a light source (in our case, an LSC transmitting sunlight) with a standard illuminant operating at the same correlated color temperature (CCT). It does so by looking at the difference in perceived color of fourteen standard samples (listed in Table 6.2) when illuminated by the tested light source to that of the standard illuminant. This difference gets assigned a value, with 100 being a perfect match with the reference, and mismatches being lower, and sometimes even negative values. The average of the first eight of these values is the color rendering index. The European Standard (EN 12646:2011) requires a CCT between 3000 K and 5300 K (warm and intermediate white light) and a minimal R_a of 80 for work spaces like offices.

The perceived color of an LSC can be described by another CIE standard: the tristimulus values X, Y, Z. These values correlate the spectrum of the illuminant of interest with the spectral sensitivity of cone cells in human eyes, and can straightforwardly be found by integrating the transmitted spectrum $T(\lambda)S(\lambda)$ multiplied by the cone responsivity $\bar{x}(\lambda)$, $\bar{y}(\lambda)$, or $\bar{z}(\lambda)$ over the visible spectrum λ_{vis}.

$$X = K \int_{\lambda vis} T(\lambda)S(\lambda)\bar{x}(\lambda)\,\mathrm{d}\lambda, \tag{6.39}$$

for X, with K a scaling factor between 0 and 100. For Y and Z similar formulas apply.

Table 6.2 List of test color samples.

Name	Approximate notation Munsell	Appearance under daylight	Swatch
TC S01	7,5 R 6/4	Light greyish red	
TC S02	5 Y 6/4	Dark greyish yellow	
TC S03	5 GY 6/8	Strong yellow green	
TC S04	2,5 G 6/6	Moderate yellowish green	
TC S05	10 BG 6/4	Light bluish green	
TC S06	5 PB 6/8	Light blue	
TC S07	2,5 P 6/8	Light violet	
TC S08	10 P 6/8	Light reddish purple	
TC S09	4,5 R 4/13	Strong red	
TC S10	5 Y 8/10	Strong yellow	
TC S11	4,5 G 5/8	Strong green	
TC S12	3 PB 3/11	Strong blue	
TC S13	5 YR 8/4	Light yellowish pink	
TC S14	5 GY 4/4	Moderate olive green (leaf)	

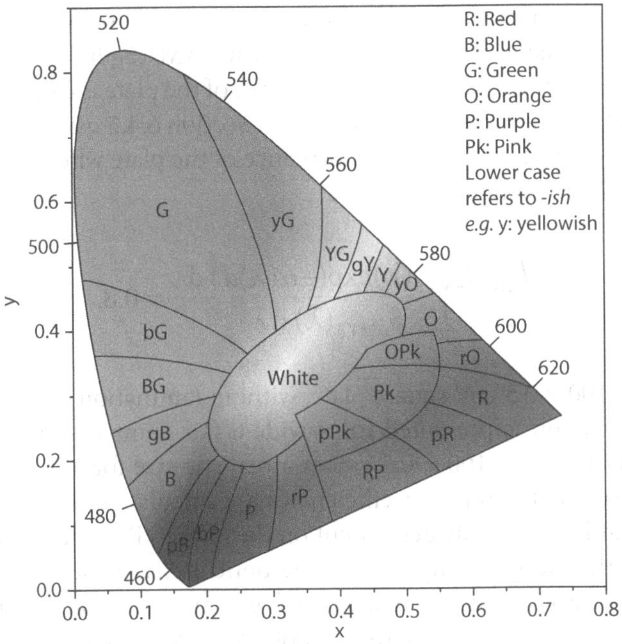

Figure 6.13 Example of the CIE 1931 color space chromaticity diagram with approximate regions and descriptions for colors added. The curve enclosing the colored regions indicates monochromatic light with wavelengths as indicated along the curve (Adapted from [63]).

The tristiniulus values can be normalized to chroniaticity color coordinates *x*, *y* and *z*, e.g.:

$$x = \frac{X}{X+Y+Z}. \tag{6.40}$$

Here, *x* and *y* represent the chromaticity, and *Y* a measure for luminance. These color coordinates make it possible to compare colors and can be plotted in a standardized CIE 1931 color space chromaticity diagram, an illustration of which is shown in Figure 6.13.

6.5.2 Evaluating the Performance of State-of-the-Art LSCs

To compare developments in LSCs, we have compiled a list of recently published materials and "classical" LSC materials, such as BASF Lumogen F

Red 305, and compared them based on a simulated 100 × 100 × 0.5 cm³ plate. This simulated plate is illuminated with AM1.5g light incident normal to the face of the plate. The performance of the plate is calculated using the Monte Carlo approach, as described in Section 6.4.5 using 10^6 photons. An assessment is made on the performance of the plate when transmitting 80% of the incident light, defined as

$$\frac{\int_{\lambda vis} S_{AM1.5}(\lambda)\exp(-\alpha(\lambda)d)\, d\lambda}{\int_{\lambda vis} S_{AM1.5}(\lambda)\, d\lambda} = 0.8. \tag{6.41}$$

A 100 × 100 × 0.5 cm³ square plate, without lamination on glass or tandem construction is presented to provide a fair comparison between all LSC materials. 100 × 100 × 0.5 cm³ is a realistic size for a window in the built environment. No power efficiencies are calculated, since the performance of an LSC is highly dependent on the chosen PV cell. A fairer comparison is therefore to only look at the optical efficiencies for each LSC material. For each family of materials: dyes, rare earth ions, and (doped) quantum dots, an overview of the photoluminescent quantum yield, CCT, and R_a at 80% transmission can be found in their respective sections. An overview of absorptions and photoluminescent emission, combined with the perceived color at 80% transmission can be seen in Figures 6.14 to 6.18,

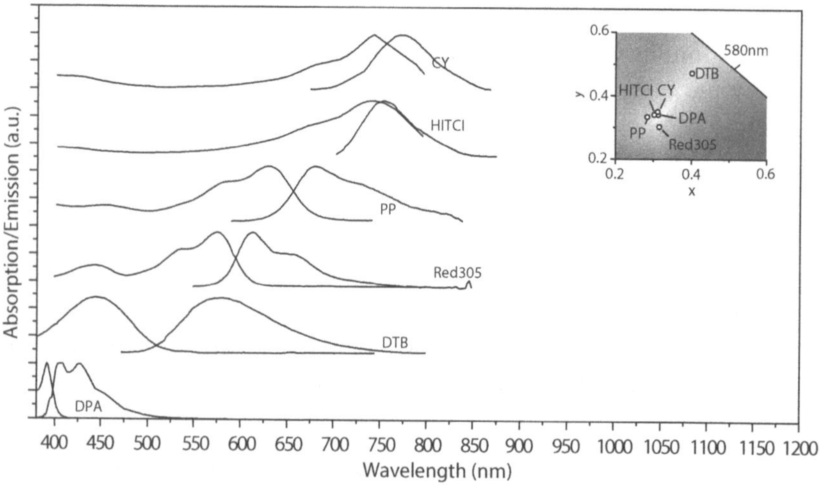

Figure 6.14 Absorption and emission spectra of dye materials used for LSCs. In the inset the color coordinates in a zoomed-in CIE1931 diagram of the dyes at 80% visible light transmission are shown.

also located in those sections. All materials are assumed to be evenly dispersed in PMMA plates (n = 1.49), with no absorption from the matrix material. The calculated efficiencies in the following section are therefore an upper limit for a realistic LSC plate. Setting these constraints allows for comparison of all active compounds, not limited by fabrication methods.

6.5.3 Dye-Based Luminescent Solar Concentrators

The dyes presented here are amongst the more "classical" LSC materials, and therefore also the most mature. The overview as presented in Figure 6.14 is far from all available dyes, as many exist (such as the Lumogen series from BASF, or the Fluorophores from Thermo Fisher [66]). Therefore, we have chosen to only present "proven" dyes, used in large-scale experiments (BASF Lumogen F Red 305 [Red305], diphenylanthracene [DPA], di-benzo-thiadiazole [DTB] [64]), or recently published dyes with high experimental significance (perylene-perinone [PP] [7], and two cyanide derivatives: CY, HITCI [65]).

As can be seen in Table 6.3, the quantum yields of these classical dyes are excellent for luminescent solar concentrators, with Red305 repeatedly reported with achieving 100% quantum yield. These dyes do, however, have their drawbacks. At 80% visible transmission, these dyes have a faithful color reproduction but are inefficient; with the UV and blue absorbing DPA only achieving 0.2% optical efficiency. As seen in Figure 6.14, DPA's coloration is low, with a high color rendering index (>90, see Table 6.3), centered around the "white light" locus in the CIE 1931 chromaticity

Table 6.3 Reported quantum yields and simulation results for 80% visible light transmission in dye-based LSCs. Both the average R_a and the worst-performing R_a are reported; 'index' specifies the name of the worst-performing color test sample (see Table 6.2).

Name	η_{QY} (%)	CCT (K)	R_a	Minimal R_a	Index	η_{opt} (%)	Ref.
DPA	83	5451	99	98	TS C10	0.2	[64]
DTB	90	3698	83	43	TS C09	2.5	[64]
PP	80	6965	96	90	TS C09	2.6	[7]
Red305	100	5077	66	-26	TS C09	3.5	[7]
CY	14	5551	96	76	TS C09	0.0	[65]
HITCI	28	5965	96	75	TS C09	0.0	[65]

diagram (see Figure 6.13). Alternatively, the dyes have an excellent optical efficiency, with DTB at 2.5% and Red305 at 2.1%, but suffer from heavy yellow and red coloration, respectively. The high efficiency dyes are therefore used in architecture applications where continued presence is not required. For instance, Red305 was used as a luminescent solar concentrator noise barrier next to a highway [67], and a mixture of DPA and DTB has been utilized as a roof to charge electric bikes [64, 68].

A newer class of dye materials (CY and HITCI) aims to overcome the heavy coloration by shifting the absorption from the visible spectrum to the infrared. Unfortunately, these materials also face some challenges, as their small Stokes shift leads to much reabsorption, with the low quantum yields not able to compensate for these reabsorption effects.

6.5.4 Rare Earth-Based Luminescent Solar Concentrators

As seen in Figure 6.15, the europium-doped $Eu(TTA)_3(TPPO)_2$ is a clear example of the benefits and drawbacks of using UV absorbing rare earth organometallic complexes in LSCs. Due to the large shift in energy[11] between the absorption by the $(TTA)_3(TPPO)_2^{3-}$ dye ligand, followed by well-defined f-f emission at 614 nm, no self-absorption is observed while retaining an excellent color rendering index of minimally 97 (see Table 6.4). Combined with the high quantum yield, nearly all absorbed photons will be emitted, contributing to the total device efficiency. Unfortunately, only a small fraction of all solar photons are available in the UV part of the solar spectrum, placing a serious cap on the efficiency. At 20% visible light absorption, only 1.3% of incident photons are absorbed. Combined with a trapping efficiency of 74%, this yields a maximally achievable η_{opt} of 0.7%.

A notable exception to the materials reported in this chapter are the Yb^{3+}-doped $CsPbCl_3$ perovskite nanocrystals [70]. This material belongs to a new class of LSC materials that try to increase the quantum yield for an LSC beyond 100% by employing quantum cutting. Quantum cutting is a phenomenon where one high energy photon (ideally) gets converted to two low energy photons by exploiting energy transfer within a material. This approach allows for an LSC that only absorbs UV photons, and is therefore transparent in visible light. The absorbed sunlight can then be emitted as infrared photons that do not suffer from reabsorption. Unfortunately, currently quantum cutting only seems to work at low photon fluencies for Yb^{3+}-doped $CsPbCl_3$ perovskite nanocrystals, with quantum yields

[11]Not to be confused with Stokes shift, which only applies to a transition within the same electronic state.

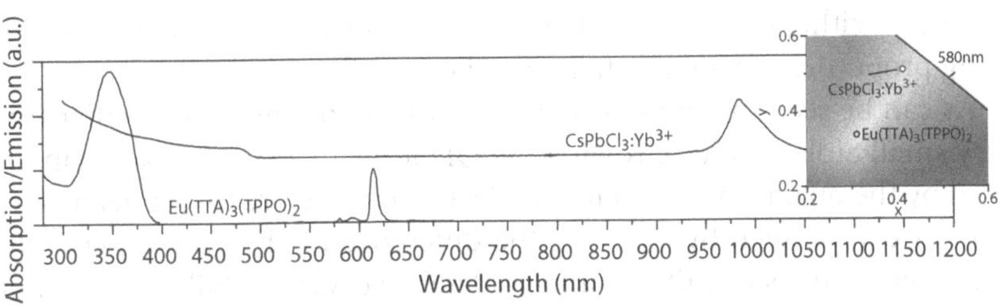

Figure 6.15 Absorption and emission spectra of rare earth-based LSCs. In the inset the color coordinates in a zoomed-in CIE1931 diagram of the rare earths at 80% visible light transmission are shown.

Table 6.4 Reported quantum yields and simulation results of 80% transmission of visible light LSC based on rare earth ions. Both the average R_a and the worst-performing R_a are reported; 'index' specifies the name of the worst-performing color test sample (see Table 6.2).

Name	η_{QY} (%)	CCT (K)	R_a	Minimal R_a	index	η_{opt} (%)	Ref.
$Eu(TTA)_3(TPPO)_2$	73	5492	99	97	TC S12	0.7	[69]
$CsPbCl_3:Yb^{3+}$	30[a]	3710	72	42	TC S09	1.5	[70]

[a]200% at extremely low photon incidence.

diminishing when exposed to high powers. This effect leads to a quantum yield of 30% when exposed to sunlight [71].

6.5.5 Quantum Dot and Doped Quantum Dot-Based Luminescent Solar Concentrators

The sheer size of the figures and tables presented in this section already indicates the popularity of quantum dots for LSC research. The tunable emission and absorption which, to a first degree, only depend on the size and chosen material of the dot, make QDs a versatile research direction. At the same time, QDs are cheap to fabricate and easily dispersed in organic waveguide materials, either in polycarbonate plates or in laminated layers, which makes scaling up to LSC prototypes possible while the material itself is still in the lab stage. A recent development for quantum dot LSCs are attempts to free the dots of toxic materials such as cadmium or lead. This is done by either choosing a different crystal, such as CuInSe or AlS, or by using properties of nanocrystalline carbon or silicon.

As can be seen in Figure 6.16, QDs made up of only a core suffer from very small Stokes shifts. Growing the QD leads to a smaller bandgap dictated by the quantum confinement effect, and therefore to a larger absorption, as can be seen by studying the difference in absorption between 3 and 4 monolayers of CdSe as a nanoplatelet versus its "full" quantum dot form. Combined with the low quantum yield of these QDs, the optical efficiency of direct bandgap QDs does not exceed a few tenths of a percent, as seen in Table 6.5. An exception to what is observed for core QDs are QDs composed solely from Si. Figure 6.16 shows that these Si QDs have almost no overlap between their emission and absorption. This effect can be attributed to the luminescence coming from the indirect bandgap of the Si QDs. This luminescence from the indirect bandgap is more prevalent than the quantum confinement effect [77]. This smaller self-absorption in Si leads to an optical efficiency of 1.3%, while still having a fair color rendering index, carrying mild yellow coloring (see Figure 6.16).

In the core/shell QDs presented here, the shell acts as sensitizer, and the core as activator. The much broadened absorption provided by the shell leads to significantly improved color rendering over their core-only varieties. Furthermore, these QDs are typically reported with reasonable

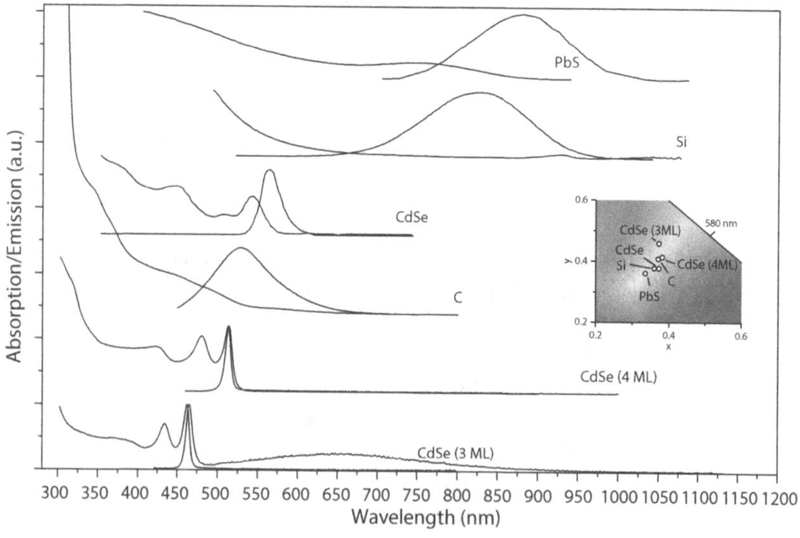

Figure 6.16 Absorption and emission spectra of LSCs based on core quantum dots. In the inset the color coordinates in a zoomed-in CIE1931 diagram of the QDs at 80% visible light transmission are shown.

quantum yields of over 50%. This combination leads to the core/shell quantum dots being amongst the most efficient LSC materials.

Notable systems are the $CdInSe_2/ZnS$, $CuInS_2/ZnS$ and AlS/ZnS LSCs. For $CdInSe_2/ZnS$, a 6.4% optical efficiency was shown in a 15.2 × 15.2 cm^2 device. These would yield a 2.4% optical efficiency at 80% transmission, without inducing artificial color blindness. The $CuInS_2/ZnS$ quantum dots, for which an optical efficiency of 8.1% was demonstrated for a 10 × 10 cm^2 LSC [81], lead in our simulations to an 2.8% efficiency at 80% transmission while still boasting an excellent R_a, as seen in Table 6.6. A similar story applies to the AlS/ZnS QDs, where, still with excellent color rendering, high efficiencies of 2.7% could be reached. What makes these last two systems notable is that they retain a high efficiency, without having to resort to toxic lead or cadmium.

The luminescence of the core QDs can be improved by doping the dots with transition metals, such as copper. For CdSe, doping with copper leads to the core of the QD acting as an activator, transferring the absorbed energy to the Cu doping, which will then emit at the absorbed energy with a higher quantum yield than the sole core would, while at the same time shifting the emission out of the self-absorption region. The effect of this for CdSe nanoplatelets can be observed in Table 6.7. Doping with copper bumps the efficiency significantly, from 0.1% to 2.4%. Unfortunately,

Table 6.5 Overview of simulation results for core quantum dots as well as reported quantum yields. Both the average R_a and the worst-performing R_a are reported, 'index' specifies the name of the worst-performing color test sample (see Table 6.2).

Name	η_{QY} (%)	CCT (K)	R_a	Minimal R_a	Index	η_{opt} (%)	Ref.
PbS	30	4638	99	96	TC S12	0.8	[72]
Si	46	3991	97	87	TC S09	1.3	[73]
CdSe	50[a]	3615	91	77	TC S12	0.3	[74]
C	30	3911	93	75	TC S09	0.1	[75]
CdSe (4 ML)[b]	5.8	4123	81	59	TC S09	0.1	[76]
CdSe (3 ML)[c]	24.3	3730	85	53	TC S09	0.1	[76]

[a]Quantum yield taken from Ref. [72].
[b]Nanoplatelets, 4 monolayers (ML) in thickness.
[c]Nanoplatelets, 3 ML in thickness.

Table 6.6 Overview of simulation results for core/shell quantum dots as well as reported quantum yields. Both the average R_a and the worst-performing R_a are reported; 'index' specifies the name of the worst-performing color test sample (see Table 6.2).

Name	η_{QY} (%)	CCT (K)	R_a	Minimal R_a	Index	η_{opt} (%)	Ref.
PbS/CdS (0.5-0.7 nm)	40	5170	95	81	TC S09	1.9	[78]
PbS/CdS (0.1-0.2 nm)	50	5419	97	86	TC S09	1.6	[78]
CISeS/ZnS	40	4803	98	94	TC S12	1.3	[79]
AlS/ZnS	60.3	4362	99	94	TC S12	2.7	[80]
$CuInS_2$/ZnS[a]	91(66)	4868	98	95	TC S12	4.0(2.8)	[81]
$CdInSe_2$/ZnS	78	4499	99	96	TC S12	2.8	[11]
$CuInS_2$/CdS	86	4704	98	92	TC S09	3.6	[82]
CdSe/CdS (14 ML)	45	3673	86	51	TC S09	1	[83]
CdSe/CdS (19 ML)	40	3698	81	47	TC S09	1.6	[80]

[a]Reports both values for solution and when dispersed in a waveguide.

Table 6.7 Overview of simulation results for doped core and doped core/shell quantum dots. Both the average R_a and the worst-performing R_a are reported; 'index' specifies the name of the worst-performing color test sample (see Table 6.2).

Name	η_{QY} (%)	CCT (K)	R_a	Minimal R_a	Index	η_{opt} (%)	Ref.
CdZnS/ZnS:Mn^{2+}	75	3864	85	67	TC S09	1.5	[11]
ZnSe/ZnS:Mn^{2+}	53	3685	81	45	TC S09	2.5	[29]
CdSe:Cu^{+}	40	4485	99	94	TC S09	1.7	[84]
CdSe (3 ML):Cu^{+}	80	4001	84	63	TC S09	1.2	[76]
CdSe (4 ML):Cu^{+}	97	3810	86	58	TC S09	2.4	[76]

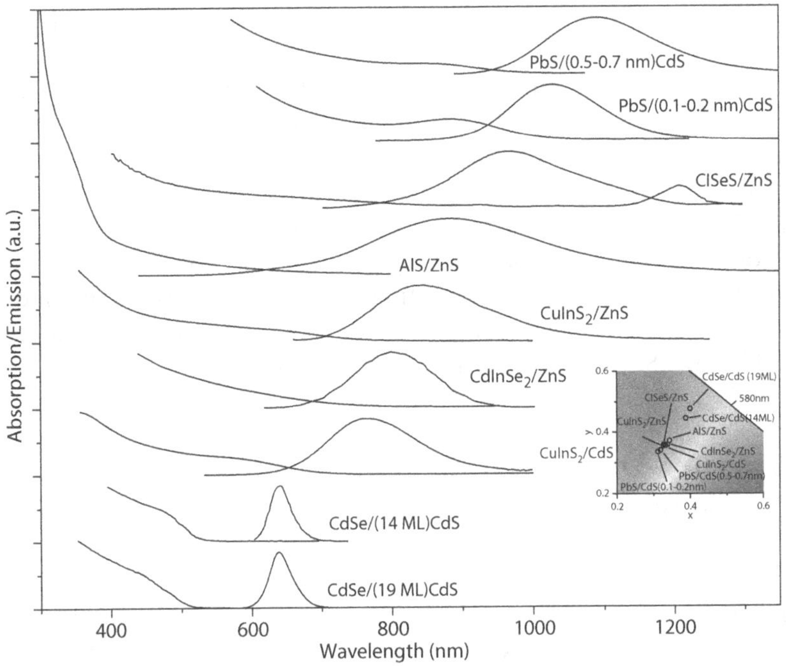

Figure 6.17 Absorption and emission spectra LSCs based on core/shell quantum dots. In the inset the color coordinates in a zoomed-in CIE1931 diagram of the QDs at 80% visible light transmission are shown.

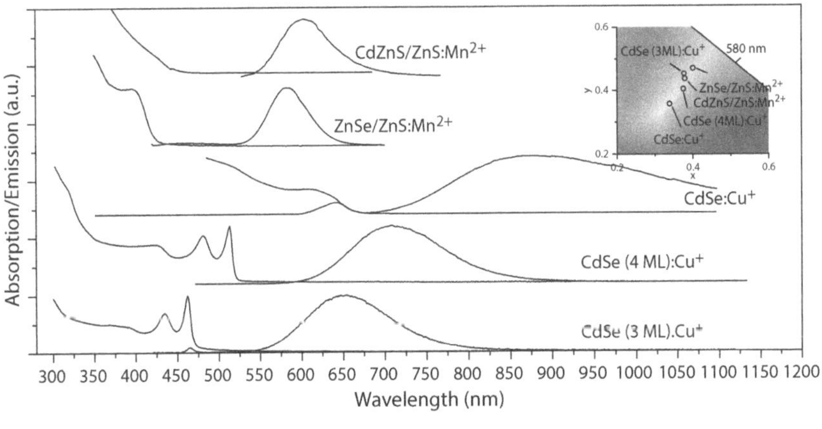

Figure 6.18 Absorption and emission spectra LSCs based on doped core and doped core/shell QDs. In the inset the color coordinates in a zoomed-in CIE1931 diagram of the QDs at 80% visible light transmission are shown.

the CdSe still dictates the absorption, which leads to a low R_a and yellow coloring.

Similar to the core-only QDs, core/shell QDs can be doped with transition metals to boost their efficiency, and remove overlap between absorption and emission. While the doped materials in literature do not show very broad absorption, limiting just over 400 nm for both Mn^{2+}-doped CdZnS/ZnS and ZnSe/ZnS, the lack of self-absorption still puts them near the top performers, demonstrating good scalability for practical applications. However, the limited absorption does yield color rendering indices that are only just sufficient for application in an office environment, when considering LSCs with 80% visible light transmission.

6.6 Tm^{2+}-Doped Halide Luminescent Solar Concentrators

A recently emerged class of possible LSC materials is the family of alkali halide salts, doped with the rare earth element thulium, in its 2+ valence state [43, 85]. As can be seen in Figure 6.19, it sets itself apart from most other contemporary LSC materials by having a very broad absorption spectrum, covering the entire visible spectrum. In the case of CaI_2, Tm^{2+} even absorbs 63% of the entire solar spectrum. What also makes thulium-doped halides unique amongst other LSC materials is the characteristic emission

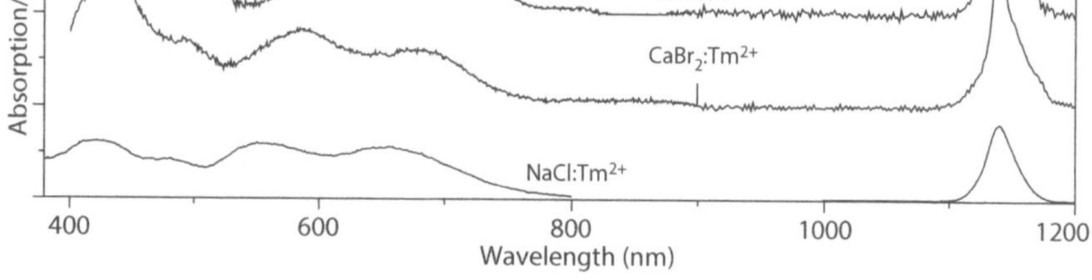

Figure 6.19 Absorption and emission of several Tm^{2+}-doped halides. In the inset the color coordinates in a zoomed-in CIE1931 diagram of the Tm^{2+}-doped halides at 80% visible light transmission are shown.

of Tm. As explained in Section 6.2.2, divalent thulium will emit most strongly at a single line centered around 1140 nm regardless of what host it is doped in.

The broad absorption observed in Tm^{2+}-doped halides is attributable to the partially filled f-shell of Tm^{2+}, occupied by thirteen electrons. All these electrons lead to a system with many possible transitions to the unoccupied 5d shell. These transitions have such low energies while being great in number that they are able to cover the entire visible spectrum. As an added advantage, because of the proximity of the lowest 5d level to the highest 4f level (in terms of energy), an excited Tm^{2+} cation can efficiently non-radiatively relax to the highest 4f level, whereupon it will emit the characteristic 1140nm light resultant from a $4f^{13}[^2F_{5/2}] \rightarrow 4f^{13}[^2F_{7/2}]$ transition. This all leads to a material with a difference between emission and absorption that is sufficiently large for Tm^{2+}-doped halides to have negligible self-absorption, as seen in Figure 6.19.

The divalent valence state for Tm^{2+} is not often encountered, due to the groundstate level of Tm often lying close to or in the conduction band. For a rare earth ion to be stable in its 2+ state, the distance between the Fermi level and the lowest 4f level ($E_{Ff} = E_{4f} - E_{Fermi}$) should ideally be smaller than 0 [87]. Due to the high lying 4f level of divalent Tm compared to other rare earth elements, this condition is not often reached. The halides, as presented here, are an exception, because of their low-lying valence band combined with a large bandgap. This combination leads to the Fermi level being sufficiently near the lowest 4f level for the $E_{Ff} \lesssim 0$ condition to be fulfilled.

Next to their broad absorption, Tm^{2+}-doped halides are also known for their remarkably high absorption coefficient. Research by Grimm, Beurer, and Güdel [88] shows (decadic) absorptions of ~ 60 at.%$^{-1}$cm^{-1} for Tm^{2+}-doped $CsCaCl_3$. Those materials were synthesized as single crystals using the Bridgman technique. While those measurements are done on bulk crystals, something still can be said about how their properties would translate to an LSC device. For most rare earth materials, concentration quenching starts at around 1% of doping. This means that, to absorb 20% of the incoming light, a thin film of 16 μm thickness would be required. To put those numbers into perspective: the concentration of Lumogen F Red 305 in PMMA can be increased to an absorption coefficient of 30 cm^{-1} before concentration quenching starts occurring.[12] This means that for a

[12]Desmet *et al.* [7] report 12 cm-1 decadic absorption for Red305 in PMMA at a concentration of 115 ppm. With PMMA having a molecular weight Mw = 120000, this leads to an intermolecular separation of 11.4 nm. Quenching starts at ~ 8nm separation [6], leading to an absorption of 30 cm-1.

Red 305 LSC, 600 μm of material is required for 20% absorption. In the case of Tm-doped halides, this opens the door to produce LSCs using physical vapor deposition [85, 86, 89], as is commonly used in the glass industry for the application of, for example, low-e coatings, or anti-reflection layers. All materials presented in this section have already been successfully deposited as thin films [85, 86].

A serious constraint on the successful development of a Tm-based LSC is the hygroscopic nature of the halides. Apart from NaCl, all other halides react with and dissolve in regular, even indoor, air. A halide LSC is possible, but an area for development therefore also has to be enclosing the luminescent coating in a protective atmosphere. An important consideration for these environments is that, contrary to how regular window glass is sealed, the sealing should be transparent to 1140 nm light. Otherwise, the sealing will interfere with the workings of the LSC.

An estimation of the potential of Tm-based LSCs can be easily made, as these doped halides do not display any self-absorption. Table 6.8 presents an overview of simulated properties of recently made halides doped with Tm^{2+} using the same simulation conditions as in the previous sections. All halides still present good color rendering indices, with $CaBr_2:Tm^{2+}$ performing best. $CaBr_2:Tm^{2+}$ shows almost no coloration, boasting a color rendering index higher than 94, with its color close to the white locus, seen in the inset of Figure 6.19. These simulations assume 100% quantum yield, leading to optical efficiencies of ~ 4.5% for all halides presented. Because of the lack of self-absorption, a lower quantum yield can easily be taken into account with $\eta_{opt} = \eta_{opt,QY=1} \times \eta_{QY}$. Even at low quantum yields, the lack of self-absorption would mean the Tm-based LSCs could outperform

Table 6.8 Properties of Tm^{2+} halide-based LSCs, at 80% transmission of visible light. For these materials, the same assumptions are made as for the simulations presented in Section 6.5. Both the average R_a and the worst-performing R_a are reported; 'index' specifies the name of the worst-performing color test sample (see Table 6.2).

Name	CCT (K)	R_a	Minimal R_a	Index	η_{opt} (%)	Ref.
$CaI_2:Tm^{2+}$	4755	95	67	TC S09	4.5	[86]
$NaI:Tm^{2+}$	5274	91	76	TC S09	4.5	[86]
$CaBr_2:Tm^{2+}$	5276	98	94	TC S09	4.5	[86]
$NaCl:Tm^{2+}$	5601	95	88	TC S10	4.4	[43]

contemporary LSC materials. Based on work with Eu^{2+}, it is not expected that rare-earth-doped halides have a low quantum yield. Eu^{2+} has photoluminescent quantum yields at 100% when doped in NaCl and NaBr [90]. It is therefore expected that Tm^{2+}-doped halides will behave similarly and will therefore make an efficient LSC.

6.7 LSC for an IPV Perspective

6.7.1 Performance Assessment

Until now, the focus of this chapter has mainly been on the use of LSCs at the borderline between in- and outdoor photovoltaics. Operating as a window, an LSC receives most of its energy from sunlight incident on the outside of a building, but it also receives energy from illumination used on the inside. As outlined before, transparency and high color rendering indices are of paramount importance for application as a window. However, it is also possible to utilize LSCs as a design element embedded in furniture or Internet of Things (IoT) devices. For these applications, the color rendering index becomes of minor importance, and adding color to a design might even be seen as beneficial [91]. Furthermore, LSCs can be bent and molded into various shapes without much impact on performance [73].

Calculating the performance of an LSC used for indoor applications happens in much the same way as for windowing applications, explained

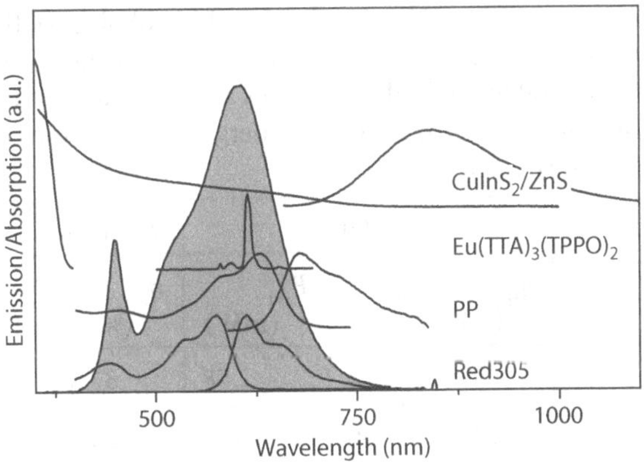

Figure 6.20 Absorption and emission spectra of selected LSC materials from Section 6.5 superimposed on the emission spectrum of an Osram SOLERIQ® S white LED (gray filled) [92].

in Section 6.4. Instead of 1000 W/m^2 from AM1.5g sunlight, for indoor applications a white LED (Figure 6.20) can be used with typical IPV powers of 1.5 W/m^2 [93]. Since indoors only visible and IR light is available, a UV-absorbing LSC will not be of any use as IPV. Taking the idealized visible light-absorbing LSC from Section 6.4.2 and performing the same calculations, but this time with the white LED as input spectrum, yields an optical efficiency of 29% to 52%, for an LSC the size of a small tabletop (55 × 55 × 2 cm^3) at 50% to 90% absorption. These optical efficiencies translate to a power output of 276 mW to 495 mW, based on the energy of the output photons. Such powers are enough to power a low-power IoT device (LoRa) or a Bluetooth antenna without the need for batteries [93].

Moving to a more realistic example, it is good to look at the LSC materials from the previous sections that absorb visible light. The top nontoxic performers from the previous sections—the Red305 dye and $CdInSe_2$/ZnS QD—have this characteristic. We can also include the PP dye, because of its good match with the white LED spectrum. As can be seen in Figure 6.20, rare-earth-doped $Eu(TTA)_3(TPPO)_2$ absorbs only UV light, and will therefore not show any appreciable power output.

We can run the same MC code as in Section 6.5, but now with a 1.5 W/m^2 white LED as input. Here we use a 55 × 55 × 2 cm^3-doped PMMA slab with either 50% or 10% transmission of the incident LED light; 50% transmission is chosen to maintain the design characteristic of LSCs, while 10% would, barring self-absorption, yield best performance. These calculations yield the results displayed in Table 6.9.

Table 6.9 Performance of 55 × 55 × 2cm^3-doped PMMA slabs of the best performing LSC materials from Section 6.5 when illuminated by white LED light giving off 1.5 W/m^2 (500 lux) of light. Materials are simulated with 50% and 10% transmission. Output power is based on the energy of the photons exiting the LSC, not on PV performance.

		T = 50%		**T = 10%**		
Name	η_{QY} (%)	η_{opt} (%)	P_{out} (mW)	η_{opt} (%)	P_{out} (mW)	**Ref.**
PP	80	22	78	37	127	[7]
Red305	100	30	115	20	70	[7]
$Eu(TTA)_3(TPPO)_2$	73	0	0	0	0	[69]
$CdInSe_2$/ZnS	78	29	87	53	153	[81]

From Table 6.9 we can see that an unobstructed LSC used as tabletop can give reasonable performance: about a third of the performance of our model system. In Red305, the effect of parasitic self-absorption can be noted when decreasing the transparency of the device. Here, the performance drops from a power output of 115 mW at 50% transparency to 70 mW at 10%. All other materials see their performance increase as transparency decreases. Here it is again the case that a small LSC tabletop would be able to power a Bluetooth antenna, or a low-power IoT device. What can also be noted when comparing the results displayed in Table 6.9 to the results from Section 6.5 are the high optical efficiencies. These high numbers can be attributed to the excellent match between the white LED emission and the LSCs absorption spectra, as seen in Figure 6.20.

These high efficiencies when used indoors were also observed by Li, Sun, and Zhang [94]. They measured Red305 LSCs with high absorption (99.99% at the absorption maximum) of varying sizes (2.5 × 2.5 cm to 30.5 × 30.5 cm). For these LSCs of varying sizes, they measured power conversion efficiencies (P_{out}/P_{in}) of 8.25% to 2.64% under fluorescent lighting and 5.60% to 2.13% under LED lighting. Incandescent and halogen light performed worse with 0.23% to 0.12% and 1.34% to 0.67%, respectively, because of the large fraction of unused IR light in the spectra of those light sources. Here, the decrease in performance can also be explained by high parasitic self-absorption, due to the high optical density of their samples. It should, however, be noted that while high efficiencies were reported, the eventual power for their devices will range from 0.05 mW (at 2.5 × 2.5 cm^2) to 3 mW (at 30.5 × 30.5 cm^2) for 1.5 W/m^2 white LED input.

For IPV purposes, if transparency of the LSC is not required, other options are available to increase performance. The trapping efficiency can quickly be increased by using a diffusive back reflector behind the LSC, with an air gap between the LSC and the back reflector [95].

6.7.2 Application Examples

Apart from the work by Li *et al.* [94], no examples of systematic research of indoor-only LSCs exist at the time of writing. However, for (dye-based) LSCs as a means for design, many examples are available that show the versatility of LSCs when used for designs other than windows.

The photoluminescent "LSC paints", used by Renny *et al.* [96] are an example of creative design with LSCs. This work allows the development of aesthetically appealing LSC art by painting luminescent dyes on plastic waveguides. As can be seen in Figure 6.21a, these paints allow for creative designs, and also demonstrate the workings of LSCs in an engaging way.

Figure 6.21 Examples of colorful LSCs used as designs, (a) Waveguide with LSC material painted on under direct illumination using a UV flashlight. (b) Concept drawing of LSCs used as a tabletop for (outdoor) furniture. (c) Prototype of the "Electric Mondrian". Each panel is an LSC with PV cells covering the perimeter. Panel a is reprinted with permission from Ref. [96] (Copyright 2018 American Chemical Society). Panel b was created by Niek Eggink, Gydo Nijensteen, and Mark Zwart and is reprinted with permission from the author of Ref. [91]. Panel c is reprinted with permission from Ref. [97] (Copyright 2017 WILEY-VCH Verlag GmbH & Co. KGaA).

The paints were therefore also developed as a means to educate students about how to harness the energy of the sun, and also the working principle and potential of LSCs.

While only being conceptual works, the demonstrations from a student project by Reinders *et al.* [91] also show the versatility of LSCs outside of windowing applications. In their article, they demonstrate how in the hands of designers, LSCs can be used to make all sorts of furniture (Figure 6.21b) or even entire constructions like a party tent.

A final example of LSCs as design element is the "Electric Mondrian" by Sark *et al.* [97], displayed in Figure 6.21c. Inspired by the "de Stijl" artform propagated by Dutch artist Piet Mondrian, combined with art in stained glass, Sark *et al.* prototyped a faceted colorful window, based on commercially available luminescent concentrator Perspex plates and solar cells. The full-sized 1 m^2 model depicted in Figure 6.21c yields a light-to-electric conversion of 0.2%, providing 2 W energy when exposed to direct sunlight.

6.8 Conclusion

From the preceding sections can be concluded that since the conception of the luminescent solar concentrator by Weber and Lambe in the 1970s, already much has been accomplished. These days, LSCs are mainly composed of one of three classes of luminophores: dyes, QDs, and rare-earth-doped hosts. A signifier of the approaching maturity of the LSC concept is that industry is starting to embrace the technique. Many materials presented in the previous sections are represented in one shape or form in industrial products. The dyes from Ref. [65] have led to the company Ubiquitous Energy [98]. The quantum dots from Ref. [73] laid the basis for Glass to Power [99] and those presented in Refs. [79, 81] to the founding of UbiQD [100]. As for the thulium-based LSCs, here industrial interest is also sparked, with the NaCl : Tm^{2+} thin films leading to the founding of Physee [101].

As the LSC concept is making its way out of the lab and into actual windows, new challenges have started to emerge. As already touched upon in Section 6.3.4, scaling up the windows to sizes such that higher-than-unity concentration factors can be reached will lead to not only broadband and uniform absorption being required, but also a window that is free from scattering. Accomplishing these feats will still require significant efforts by the LSC community. However, as can be deduced from the previous sections, recent advances have brought us ever closer to a broadband absorbing, self-absorption free, non-scattering luminescent solar concentrator.

Acknowledgements

The authors would like to thank many people from the LSC community for contributing their data to this chapter: Dr. K. Wang, SUSTech and Dr. C. Wei, TUM, for the data from Ref. [80]; Dr. H. McDaniel and Dr. M.R. Bergren, UbiQD, for the data from Ref. [81]; Prof. Dr. F. Rosei and D. Benetti, MSc, INRS, for the data from Refs. [75, 78]; Prof. Dr. H.V. Deimir and Dr. M. Sharma, NTU, for the data from Ref. [76]; and Prof. Dr. D.R. Gamelin, T.A. Cohen, BSc, and C.S. Erickson, MSc, UW, for the data from Refs. [29, 70, 82, 84].

This work was supported by the Netherlands Organization for Scientific Research (NWO/OCW), as part of the LumiCon project (15024).

References

1. Weber, W.H. and Lambe, J., Luminescent greenhouse collector for solar radiation. *Appl. Opt.*, 15, 10, 2299–2300, 1976.
2. Batchelder, J.S., Zewail, A.H., Cole, T., Luminescent solar concentrators 2: Experimental and theoretical analysis of their possible efficiencies. *Appl. Opt.*, 20, 21, 3733, 1981.
3. Wittwer, V., Stahl, W., Goetzberger, A., Fluorescent planar concentrators. *Sol. Energy Mater.*, 11, 3, 187–197, 1984.

4. Norton, B., Eames, P.C., Mallick, T.K., Huang, M.J., McCormack, S.J., Mondol, J.D., Yohanis, Y.G., Enhancing the performance of building integrated photovoltaics. *Sol. Energy*, 85, 8, 1629–1664, 2011.
5. Debije, M.G. and Verbunt, P.P.C., Thirty years of luminescent solar concentrator research: Solar energy for the built environment. *Adv. Energy Mater.*, 2, 1, 12–35, 2012.
6. Green, A.P. and Buckley, A.R., Solid state concentration quenching of organic fluorophores in PMMA. *Phys. Chew. Chew. Phys.*, 17, 2, 1435–1440, 2015.
7. Desmet, L., Ras, A.J.M., de Boer, D.K.G., Debije, M.G., Monocrystalline silicon photovoltaic luminescent solar concentrator with 4.2% power conversion efficiency. *Opt. Letters*, 37, 15, 3087–3089, 2012.
8. Schäfer, F.P., *Dye Lasers*, F.P. Schafer (Ed.), p. 228, Springer Verlag Berlin Heidelberg, Berlin, Heidelberg, 1989.
9. Fleming, K.G., Fluorescence theory, in: *Encyclopedia of Spectroscopy and Spectrometry*, J. Lindon, G.E. Tranter, D. Koppenaal (Eds.), pp. 628–634, Academic Press, Cambridge, Massachusetts USA, 2010.
10. Slooff, L.H., Bende, E.E., Burgers, A.R., Budel, T., Pravettoni, M., Kenny, R.P., Dunlop, E.D., Biichtemann, A., A luminescent solar concentrator with 7.1% power conversion efficiency. *Phys. Status Solidi RRL*, 2, 6, 257–259, 2008.
11. Wu, K., Li, H., Klimov, V.I., Tandem luminescent solar concentrators based on engineered quantum dots. *Nat. Photonics*, 12, 2, 105–110, 2018.
12. Tummeltshammer, C., Portnoi, M., Mitchell, S.A., Lee, A.-T., Kenyon, A.J., Tabor, A.B., Papakonstantinou, I., On the ability of forster resonance energy transfer to enhance luminescent solar concentrator efficiency. *Nano Energy*, 32, 263–270, 2017.
13. Meinardi, F., Bruni, F., Brovelli, S., Luminescent solar concentrators for building-integrated photovoltaics. *Nat. Rev. Mater.*, 2, 12, 17072, 2017.
14. Haxel, G.B., Hedrick, J.B., Orris, G.J., Rare earth elements -critical resources for high technology. *U. S. Geol. Surv. USGS Fact Sheet*, 087-02, 1–11, 2002.
15. Bünzli, J.-C.G., Lanthanide photonics: Shaping the nanoworld. *Trends Chem.*, 1, 8, 1–13, 2019.
16. Blasse, G. and Grabmaier, B.C., *Luminescent Materials*, pp. 118–119, Springer Verlag, Berlin, Heidelberg, 1941.
17. Henderson, B. and Imbusch, G.F., *Optical Spectroscopy of Inorganic Solids*, Oxford University Press, New York, 1989.
18. van Sark, W.G., de Wild, J., Rath, J.K., Meijerink, A., Schropp, R.E., Upconversion in solar cells. *Nanoscale Res. Lett.*, 8, 1, 81, 2013.
19. Ronda, C., *Luminescence: From, Theory to Applications*, C. Ronda (Ed.), WILEY-VCH Verlag GmbH & Co. KGaA, Weinheim, 2008.
20. Huang, X., Han, S., Huang, W., Liu, X., Enhancing solar cell efficiency: The Search for luminescent materials as spectral converters. *Chem. Soc. Rev.*, 42, 1, 173–201, 2013.

21. van der Ende, B.M., Aarts, L., Meijerink, A., Lanthanide ions as spectral converters for solar cells. *PCCP*, 11, 47, 11081, 2009.
22. Dieke, G.H., Crosswhite, H.M., Dunn, B., Emission spectra of the doubly and triply ionized rare earths*. *J. Opt. Soc. Am.*, 51, 8, 820, 1961.
23. Ma, C.-G., Brik, M., Liu, D.-X., Feng, B., Tian, Y., Suchocki, A., Energy level schemes of f^N electronic configurations for the di-, tri-, and tetravalent lanthanides and actinides in a free state. *J. Lumin.*, 170, 369–374, 2016.
24. Norris, D.J. and Bawendi, M.G., Measurement and assignment of the size-dependent optical spectrum in CdSe quantum dots. *Phys. R. R.*, 53, 24, 16338–16346, 1996.
25. Delerue, C. and Lannoo, M., *Nanostructures - Theory and Modeling*, Springer Verlag Berlin Heidelberg, Berlin, Heidelberg, 2004.
26. Moraitis, P., Schropp, R., van Sark, W., Nanoparticles for luminescent solar concentrators - a review. *Opt. Mater.*, 84, 636–645, 2018.
27. Vasudevan, D., Gaddam, R.R., Trinchi, A., Cole, I., Core-shell quantum dots: Properties and applications. *J. Alloys Compd.*, 636, 395–404, 2015.
28. AbouElhamd, A.R., Al-Sallal, K.A., Hassan, A., Review of core/shell quantum dots technology integrated into building's glazing. *Energies*, 12, 6, 1058, 2019.
29. Erickson, C.S., Bradshaw, L.R., McDowall, S., Gilbertson, J.D., Gamelin, D.R., Patrick, D.L., Zero-reabsorption doped-nanocrystal luminescent solar concentrators. *ACS Nano*, 8, 4, 3461–3467, 2014.
30. Martin-Rodriguez, R., Geitenbeek, R., Meijerink, A., Incorporation and luminescence of Yb^{3+} in CdSe nanocrystals. *JACS*, 135, 37, 13668–13671, 2013.
31. Merkx, E.P.J., Lensvelt, T.G., van der Kolk, E., Modelling and optimization of UV absorbing photovoltaic windows using a thin film $AlN:Eu^{3+}$ luminescence library. *Sol. Energy Mater. Sol. Cells*, 200, 110032, 2019.
32. Vossen, F.M., Aarts, M.P., Debije, M.G., Visual performance of red luminescent solar concentrating windows in an office environment. *Energy Build.*, 113, 123–132, 2016.
33. Mulder, C.L., Reusswig, P.D., Beyler, A.P., Kim, H., Rotschild, C., Baldo, M.A., Dye alignment in luminescent solar concentrators: II Horizontal alignment for energy harvesting in linear polarizers. *Opt. Express*, 18, S1, A91, 2010.
34. Sol, J.A., Timmermans, G.H., van Breugel, A.J., Schenning, A.P., Debije, M.G., Multistate luminescent solar concentrator "smart" windows. *Adv. Energy Mater.*, 8, 12, 1–8, 2018.
35. Merkx, E.P.J., ten Kate, O.M., van der Kolk, E., Rapid optimization of large-scale luminescent solar concentrators: Evaluation for adoption in the built environment. *Opt. Express*, 25, 12, A547, 2017.
36. Krumer, Z., Van Sark, W. G. J. H. M., Schropp, R.E.I., De, C., Donegá, M., Compensation of self-absorption losses in luminescent solar concentrators by increasing luminophore concentration. *Sol. Energy Mater. Sol. Cells*, 167, 133–139, 2017.

37. Kastelijn, M.J., Bastiaansen, C.W., Debije, M.G., Influence of waveguide material on light emission in luminescent solar concentrators. *Opt. Mater.*, 31, 11, 1720–1722, 2009.
38. Zettl, M., Mayer, O., Klampaftis, E., Richards, B.S., Investigation of host polymers for luminescent solar concentrators. *Energy Technol.*, 5, 7, 1037–1044, 2017.
39. de Boer, D.K.G., Broer, D.J., Debije, M.G., Keur, W., Meijerink, A., Ronda, C.R., Verbunt, P.P.C., Progress in phosphors and filters for luminescent solar concentrators. *Opt. Express*, 20, 10, A395–405, 2012.
40. Rafiee, M., Chandra, S., Ahmed, H., McCormack, S.J., An overview of various configurations of luminescent solar concentrators for photovoltaic applications. *Opt. Mater.*, 91, 212–227, 2019.
41. Currie, M.J., Mapel, J.K., Heidel, T.D., Goffri, S., Baldo, M.A., High-efficiency organic solar concentrators for photovoltaics. *Science*, 321, 5886, 226–228, 2008.
42. Goldschmidt, J.C., Peters, M., Bösch, A., Helmers, H., Dimroth, F., Glunz, S.W., Willeke, G., Increasing the efficiency of fluorescent concentrator systems. *Sol. Energy Mater. Sol. Cells*, 93, 2, 176–182, 2009.
43. ten Kate, O.M., Krämer, K.W., van der Kolk, E., Efficient luminescent solar concentrators based on self-absorption free, Tm^{2+} doped halides. *Sol. Energy Mater. Sol. Cells*, 140, 115–120, 2015.
44. Freunek, M., Freunek, M., Reindl, L.M., Maximum efficiencies of indoor photovoltaic devices. *IEEE J. Photovoltaics*, 3, 1, 59–64, 2013.
45. Papakonstantinou, I. and Tummeltshammer, C., Fundamental limits of concentration in luminescent solar concentrators revised: The Effect of reabsorption and nonunity quantum yield. *Optica*, 2, 10, 841, 2015.
46. Li, H., Wu, K., Lim, J., Song, H.-J., Klimov, V.I., Doctor-blade deposition of quantum dots onto standard window glass for low-loss large-area luminescent solar concentrators. *Nat. Energy*, 1, 12, 16157, 2016.
47. Yoshikawa, K., Kawasaki, H., Yoshida, W., Irie, T., Konishi, K., Nakano, K., Uto, T., Adachi, D., Kanematsu, M., Uzu, H., Yamamoto, K., Silicon heterojunction solar cell with interdigitated back contacts for a photoconversion efficiency over 26%. *Nat. Energy*, 2, 5, 17032, 2017.
48. Green, M.A., Dunlop, E.D., Levi, D.H., Hohl-Ebinger, J., Yoshita, M., Ho-Baillie, A.W., Solar cell efficiency tables (version 54). *Prog. Photovoltaics Res. Appl.*, 27, 7, 565–575, 2019.
49. Klimov, V.I., Baker, T.A., Lim, J., Velizhanin, K.A., McDaniel, H., Quality factor of luminescent solar concentrators and practical concentration limits attainable with semiconductor quantum dots. *ACS Photonics*, 3, 6, 1138–1148, 2016.
50. Kate, O.M., Hooning, K.M., van der Kolk, E., Quantifying self-absorption losses in luminescent solar concentrators. *Appl. Opt.*, 53, 23, 5238, 2014.
51. Merkx, E.P.J., ten Kate, O.M., van der Kolk, E., MGLT model in MATLAB, 2020. https://github.com/emerkx/multiple-generation-light-tranport.

52. Krumer, Z., Pera, S.J., van Dijk-Moes, R.J., Zhao, Y., de Brouwer, A.F., Groeneveld, E., Sark, W.G., van, Schropp, R.E., de Mello Donegá, C., Tackling self-absorption in luminescent solar concentrators with type-II colloidal quantum dots. *Sol. Energy Mater. Sol. Cells*, 111, 57–65, 2013.
53. Freunek-Miiller, M. (Ed.), *Photovoltaic Modeling Handbook*, John Wiley & Sons, Inc., Hoboken, NJ, USA, 2018.
54. Synopsys, Lighttools. https://www.synopsys.com/optical-solutions/lighttools.html.
55. COMSOL Inc, COMSOL Multiphysics®, https://www.comsol.com.
56. Farrell, D.J., Poelking, C., Godel, K.C., Fuhrer, M., Chatten, A., 2019. Pvtrace, https://github.com/danieljfarrell/pvtrace.
57. Merkx, E.P.J., Monte Carlo MATLAB example, 2020. https://github.com/emerkx/lsc-monte-carlo.
58. Sansregret, J., Drake, J.M., Thomas, W.R.L., Lesiecki, M.L., Light transport in planar luminescent solar concentrators: The Role of dcm self-absorption. *Appl. Opt.*, 22, 4, 573–577, 1983.
59. Şahin, D., Ilan, B., Kelley, D.F., Monte-carlo simulations of light propagation in luminescent solar concentrators based on semiconductor nanoparticles. *J. Appl. Phys.*, 110, 3, 033108, 2011.
60. Byrnes, S.J., Multilayer optical calculations, Retrieved from https://arxiv.org/abs/1603.02720, 2016.
61. Gutmann, J., Peters, M., Bläsi, B., Hermle, M., Gombert, A., Zappe, H., Goldschmidt, J.C., Electromagnetic simulations of a photonic luminescent solar concentrator. *Opt. Express*, 20, S2, A157, 2012.
62. International Commission on Illumination, in: *Method of Measuring and Specifying Colour Rendering Properties of Light Sources*, CIE Central Bureau, Vienna, Austria, 1995.
63. Nave, R., The CIE chromaticity diagram, Hyperphysics, retrieved from http://hyperphysics.phy-astr.gsu.edu/hbase/vision/cie.html, 09/01/2019.
64. Aste, N., Tagliabue, L., Del Pero, C., Testa, D., Fusco, R., Performance analysis of a large-area luminescent solar concentrator module. *Renewable Energy*, 76, 330–337, 2015.
65. Zhao, Y., Meek, G., Levine, B., Lunt, R., Near infrared harvesting transparent luminescent solar concentrators. *Adv. Opt. Mater.*, 2, 7, 606–611, 2014.
66. *The Molecular Probes handbook. A guide to fluorescent probes and labeling technologies*, Life Technologies, Carlsbad, California, USA, 2010.
67. Kanellis, M., de Jong, M.M., Slooff, L., Debije, M.G., The solar noise barrier project: 1. Effect of incident light orientation on the performance of a large-scale luminescent solar concentrator noise barrier. *Renewable Energy*, 103, 647–652, 2017.
68. Conenna, N., Luminescent solar concentrator LSC, https://energycue.it/en/luminescei solar-concentrator-lsc/9655/, 09/01/2019.
69. Wang, T., Zhang, J., Ma, W., Luo, Y., Wang, L., Hu, Z., Wu, W., Wang, X., Zou, G., Zhang, Q., Luminescent solar concentrator employing rare earth complex with zero self-absorption loss. *Sol. Energy*, 85, 11, 2571–2579, 2011.

70. Cohen, T.A., Milstein, T.J., Kroupa, D.M., MacKenzie, J.D., Lus-combe, C.K., Gamelin, D.R., Quantum-cutting Yb^{3+}-doped perovskite nanocrystals for monolithic bilayer luminescent solar concentrators. *J. Mater. Chem. A*, 7, 15, 9279–9288, 2019.
71. Kroupa, D.M., Roh, J.Y., Milstein, T.J., Creutz, S.E., Gamelin, D.R., Quantum-cutting Ytterbium-doped $CsPb(Cl_{1x}Br_x)_3$ perovskite thin films with photoluminescence quantum yields over 190%. *ACS Energy Lett.*, 3, 10, 2390–2395, 2018.
72. Shcherbatyuk, G.V., Inman, R.H., Wang, C., Winston, R., Ghosh, S., Viability of using near infrared PbS quantum dots as active materials in luminescent solar concentrators. *Appl. Phys. Lett.*, 96, 19, 13–16, 2010.
73. Meinardi, F., Ehrenberg, S., Dhamo, L., Carulli, F., Mauri, M., Bruni, F., Simonutti, R., Kortshagen, U., Brovelli, S., Highly efficient luminescent solar concentrators based on earth-abundant indirect-bandgap silicon quantum dots. *Nat. Photonics*, 11, 3, 177–185, 2017.
74. Chen, Y., Vela, J., Htoon, H., Casson, J.L., Werder, D.J., Russian, D., Klimov, V.I., Hollingsworth, J.A., "Giant" multishell CdSe nanocrystal quantum dots with suppressed blinking. *JACS*, 130, 15, 5026–5027, 2008.
75. Zhou, Y., Benetti, D., Tong, X., Jin, L., Wang, Z.M., Ma, D., Zhao, H., Rosei, F., Colloidal carbon dots based highly stable luminescent solar concentrators. *Nano Energy*, 44, 378–387, 2018.
76. Sharma, M., Gungor, K., Yeltik, A., Olutas, M., Guzelturk, B., Ke-lestemur, Y., Erdem, T., Delikanli, S., McBride, J.R., Demir, H.V., Near-unity emitting copper-doped colloidal semiconductor quantum wells for luminescent solar concentrators. *Adv. Mater.*, 29, 30, 1–10, 2017.
77. Hannah, D.C., Yang, J., Podsiadlo, P., Chan, M.K., Demortière, A., Gosztola, D.J., Prakapenka, V.R., Schatz, G.C., Kortshagen, U., Schaller, R.D., On the origin of photoluminescence in silicon nanocrystals: Pressure-dependent structural and optical studies. *Nano Lett.*, 12, 8, 4200–4205, 2012.
78. Zhou, Y., Benetti, D., Fan, Z., Zhao, H., Ma, D., Govorov, A.O., Vomiero, A., Rosei, F., Near infrared, highly efficient luminescent solar concentrators. *Adv. Energy Mater.*, 6, 11, 1–8, 2016.
79. Meinardi, F., McDaniel, H., Carulli, F., Colombo, A., Velizhanin, K.A., Makarov, N.S., Simonutti, R., Klimov, V.I., Brovelli, S., Highly efficient large-area colourless luminescent solar concentrators using heavy-metal-free colloidal quantum dots. *Nat. Nanotechnol.*, 10, 10, 878–885, 2015.
80. Chen, W., Li, J., Liu, P., Liu, H., Xia, J., Li, S., Wang, D., Wu, D., Lu, W., Sun, X.W., Wang, K., Heavy metal free nanocrystals with near infrared emission applying in luminescent solar concentrator. *Solar RRL*, 1, 6, 1700041, 2017.
81. Bergren, M.R., Makarov, N.S., Ramasamy, K., Jackson, A., Guglielmetti, R., McDaniel, H., High-performance $CuInS_2$ quantum dot laminated glass luminescent solar concentrators for windows. *ACS Energy Lett.*, 3, 3, 520–525, 2018.

82. Knowles, K.E., Kilburn, T.B., Alzate, D.G., McDowall, S., Gamelin, D.R., Bright $CuInS_2$/CdS nanocrystal phosphors for high-gain full-spectrum luminescent solar concentrators. *Chem. Commun.*, 51, 44, 9129–9132, 2015.
83. Meinardi, F., Colombo, A., Velizhanin, K.A., Simonutti, R., Lorenzon, M., Beverina, L., Viswanatha, R., Klimov, V.I., Brovelli, S., Large-area luminescent solar concentrators based on 'stokes-shift-engineered' nanocrystals in a mass-polymerized PMMA matrix. *Nat. Photonics*, 8, 5, 392–399, 2014.
84. Bradshaw, L.R., Knowles, K.E., McDowall, S., Gamelin, D.R., Nanocrystals for luminescent solar concentrators. *Nano Lett.*, 15, 2, 1315–1323, 2015.
85. de Jong, M., Kesteloo, W., van der Kolk, E., Deposition of luminescent NaCl:Tm^{2+} thin films with a Tm concentration gradient using rf magnetron sputtering. *Opt. Mater.*, 46, 149–153, 2015.
86. Merkx, E.P.J. and van der Kolk, E., Optimizing Tm^{2+}-doped dihalide thin film luminescent solar concentrators, in: *OSA Advanced Photonics Congress (IPR, Networks, NOMA, SPPCom, PVLED)*, OSA Technical Digest, Optical Society of America, PT1C.3, 2019.
87. Dorenbos, P., Valence stability of lanthanide ions in inorganic compounds. *Chem. Mater.*, 17, 25, 6452–6456, 2005.
88. Grimm, J., Beurer, E., Güdel, H.U., Crystal absorption spectra in the region of 4f-4f and 4f-5d excitations in Tm^{2+}-doped $CsCaCl_3$, $CsCaBr_3$, and $CsCaI_3$. *Inorg. Chem.*, 45, 26, 10905–10908, 2006.
89. Merkx, E.P.J. and van der Kolk, E., Method for the detailed characterization of cosputtered inorganic luminescent material libraries. *ACS Comb. Sci.*, 20, 11, 595–601, 2018.
90. Rubio, J., Doubly-valent rare-earth ions in halide crystals. *J. Phys. Chem. Solids*, 52, 1, 101–174, 1991.
91. Reinders, A., Kishore, R., Slooff, L., Eggink, W., Luminescent solar concentrator photovoltaic designs. *Jpn. J. Appl. Phys.*, 57, 8S3, 08RD10, 2018.
92. OSRAM, Opto Semiconductors GmbH, *SOLERIQ S 9 Datasheet Version 1.1*, p. 8, OSRAM Opto Semiconductors GmbH, Regensburg, 2018.
93. Mathews, I., Kantareddy, S.N., Buonassisi, T., Peters, I.M., Technology and market perspective for indoor photovoltaic cells. *Joule*, 3, 6, 1415–1426, 2019.
94. Li, Y., Sun, Y., Zhang, Y., Luminescent solar concentrators performing under different light conditions. *Sol. Energy*, 188, 1248–1255, 2019.
95. Debije, M.G., Teunissen, J.P., Kastelijn, M.J., Verbunt, P.P., Bastiaansen, C.W., The effect of a scattering layer on the edge output of a luminescent solar concentrator. *Sol. Energy Mater. Sol. Cells*, 93, 8, 1345–1350, 2009.
96. Renny, A., Yang, C., Anthony, R., Lunt, R.R., Luminescent solar concentrator paintings: Connecting art and energy. *J. Chem. Educ.*, 95, 7, 1161–1166, 2018.
97. van Sark, W., Moraitis, P., Aalberts, C., Drent, M., Grasso, T., L'Ortije, Y., Visschers, M., Westra, M., Plas, R., Planje, W., The "electric mondrian" as a

luminescent solar concentrator demonstrator case study. *Solar RRL*, 1, 3–4, 1600015, 2017.
98. Ubiquitous energy, http://ubiquitous.energy/, 09/01/2019.
99. Glass to power, https://www.glasstopower.com/g2p/, 09/01/2019.
100. Ubiqd, https://ubiqd.com/, 09/01/2019.
101. Physee, https://physee.eu/, 09/01/2019.

7

Organic Photovoltaic Cells and Modules for Applications under Indoor Lighting Conditions

Birger Zimmermann[1*] **and Uli Würfel**[1,2]

[1]*Fraunhofer Institute for Solar Energy Systems ISE, Organic and Perovskite Photovoltaics, Freiburg, Germany*
[2]*Freiburg Materials Research Centre (FMF), Freiburg, Germany*

Abstract
Organic photovoltaics is based on molecular semiconductors with extended pi-electron systems expanding over the carbon backbone. This can be small to large molecules, fullerenes, oligomers and polymers. These materials offer the possibility to engineer the bandgap of the semiconductor by altering the chemical structure of the molecules. Besides changes in the color, varying from red to yellow over green and blue to purple, this also has technical implications.

For indoor applications under artificial light, the optimum bandgap is in the range of 1.9eV, matching the bandgap of a variety of OPV materials.

Organic solar cells can be produced by printing and coating in roll-to-roll processes and even by digital inkjet printing, offering fast and cost efficient production and the possibility to print nearly arbitrarily shaped solar cells. The low light intensity and related low device currents open the path to use different materials, device setups and dimensions as compared to full sun applications, where high currents demand a highly effective current collection. On the other hand it raises the requirements for defect free devices with very high values for the parallel resistance.

This chapter will discuss these opportunities and challenges from basics to the implementation of indoor optimized OPV devices.

Keywords: Organic solar cells, organic photovoltaics, OPV modules, indoor lighting

**Corresponding author*: birger.zimmermann@ise.fraunhofer.de

Monika Freunek Müller (ed.) Indoor Photovoltaics: Materials, Modeling and Applications, (189–212)

7.1 Introduction

Organic photovoltaics (OPV) is a technology where the absorber materials are carbon-based molecular semiconductors with extended pi-electron systems. These can be small to large molecules, fullerenes, oligomers and polymers. Reports on electrically conductive polymers reach back to the 1950s [1] and it was already shown that the conductivity is caused by semiconducting properties of these materials [2] and there were even indications for the possibility of doping found at that time [3]. Roughly 25 years later, significant progress was made in terms of achievable conductivities, culminating in the Nobel Prize awarded to Hideki Shirakawa, Alan G. MacDiarmid and Alan J. Heeger in the year 2000 [4]. In single material films, photo-excitation results in the formation of rather strongly bound excitons and only very few of the latter dissociate into free charge carriers at room temperature. These excitons can, however, be split at an interface to another material with almost unity quantum efficiency, which offers an energetically favorable state for either the electron or the hole (Figure 7.1, left). A photovoltaic effect for organic hetero-interfaces was already reported in 1958 [5]. Solar cells with a power conversion efficiency ≥ 1% were first reported in 1986 [6] and major improvements were achieved by the introduction of fullerene acceptors [7] and the so-called bulk heterojunction (BHJ) concept [8].

The latter—comprising an interpenetrating donor-acceptor network—allows decoupling of the average distance to a hetero-interface from the total film thickness; or, in other words, enables creating thick photoactive films with large interfacial area for exciton dissociation, as cartooned in Figure 7.1, middle. On the right-hand side of Figure 7.1 a typical device stack with respective ranges of layer thicknesses is shown. In most laboratories, indium tin oxide (ITO) is used as first electrode, while for roll-to-roll (R2R) production ITO-free solutions are highly preferred [9–11], but the use of ITO is possible as well [12].

By replacing the fullerene acceptors with molecular non-fullerene acceptors (NFA) which combine electron accepting properties with strong absorption and a high electron mobility, another significant step towards technically meaningful efficiencies was made, such that by 2019 the highest reported efficiency for single junction OPV is over 17% [13].

The molecular nature of these materials results in high absorption coefficients and offers the possibility to engineer the bandgap (E_g) of the semiconducting material by altering the chemical structure of the molecules

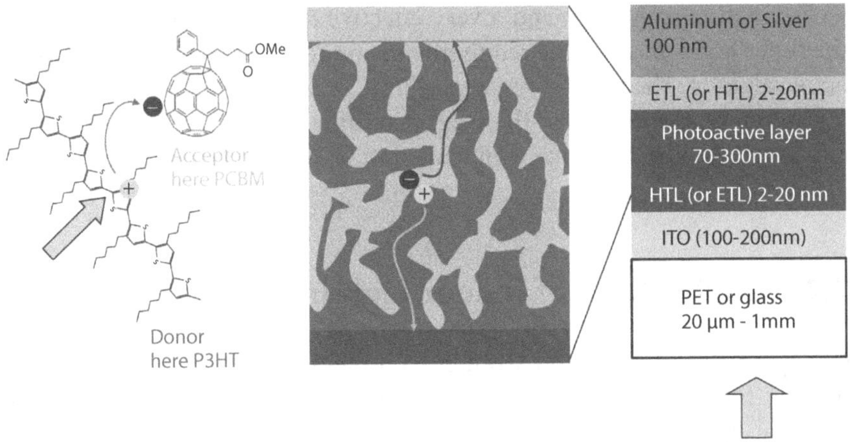

Figure 7.1 Basic principles of OPV operation and typical materials and thicknesses used.

used [14]. Besides changes in color, varying from yellow to red over green and blue to purple, this also has technical implications. Different applications demand different bandgaps for optimum performance. For indoor applications under artificial light, the optimum bandgap is in the range of 1.8–1.9 eV, as discussed in detail in Chapter 3, matching the bandgap of a variety of OPV materials [15–17].

The necessity to use a donor/acceptor system adds a significant limitation to the obtainable device efficiency (η), mainly due to a reduced open-circuit voltage (V_{OC}). It is therefore of high importance to optimize the energy level alignment and understand the exact mechanisms to achieve the highest possible V_{OC} [18].

Recent progress has allowed for devices with a V_{OC} of over 1V [19] and an estimated efficiency under 1000 lux artificial light of 26%. Another publication even claims more than 28% efficiency [20], showing the potential of the technology for highly efficient indoor photovoltaic cells. At these high efficiency values the comparability of results between laboratories and technologies should be increased, as discussed in Chapter 5.

Organic solar cells can be produced by printing and coating methods in continuous roll-to-roll processes [11, 21–24] and even by digital inkjet printing [25], offering fast and cost-efficient production and the possibility to print nearly arbitrarily shaped solar cells. The low light intensity and related low device currents open the path to use different materials, device setups and dimensions, as compared to applications under full sunlight, where high currents demand a very effective current collection [16]. On the other hand, it raises the requirements for defect-free devices with very large values for the parallel resistance [26].

This chapter will discuss these opportunities and challenges, from fundamentals to the technical implementation of optimized indoor OPV devices.

7.2 Implications of Indoor Lighting

Changing the illumination from solar radiation (AM1.5G) to artificial light in buildings has two major impacts. First, the emission spectra of artificial light sources are different from solar radiation, typically emitting much less or even no infrared light at all. All modern power-efficient light sources will emit only in the wavelength range of the sensitivity of the human eye, which roughly ends at 650–700 nm, corresponding to a bandgap of 1.8–1.9 eV. The narrow emission of modern light sources allows for much higher theoretical efficiencies due to strongly reduced thermalization losses, reaching 50–60%, as shown and discussed in detail in Chapter 2. Figure 7.2 shows the theoretical curves for AM 1.5G illumination and a fluorescent lamp as a guide to the eye, together with experimental efficiency values from the literature for Organic, Dye, and Perovskite photovoltaic (PV) cells. A non-optimized silicon device is added as an example of a narrow bandgap solar cell. The data point for the silicon solar cell and the lower efficiency OPV devices were measured by the authors under 500 lux under a fluorescent lamp (Megaman PAR38 E27 23W/840 (MM16214)). The measurement was calibrated by setting the lamp to a distance such that a Fraunhofer ISE CalLab reference solar cell showed the current value which it should give at 500 lux. This value was calculated by integrating the product of the spectral response and the calibrated lamp spectrum at 500 lux. The spectrum was calibrated by measuring the spectrum with a calibrated spectrophotometer at Fraunhofer ISE CalLab and then normalizing it such that the integral of the product of the spectrum, the photometric radiation equivalent and the human eye sensitivity function gave 500 lux. To double check for any error in the calibration procedure, the light intensity was measured with a high quality luxmeter (PCE-174, PCE Group) as a consistency check. In the majority of the literature measurements a fluorescent lamp at 1000 lux was used.

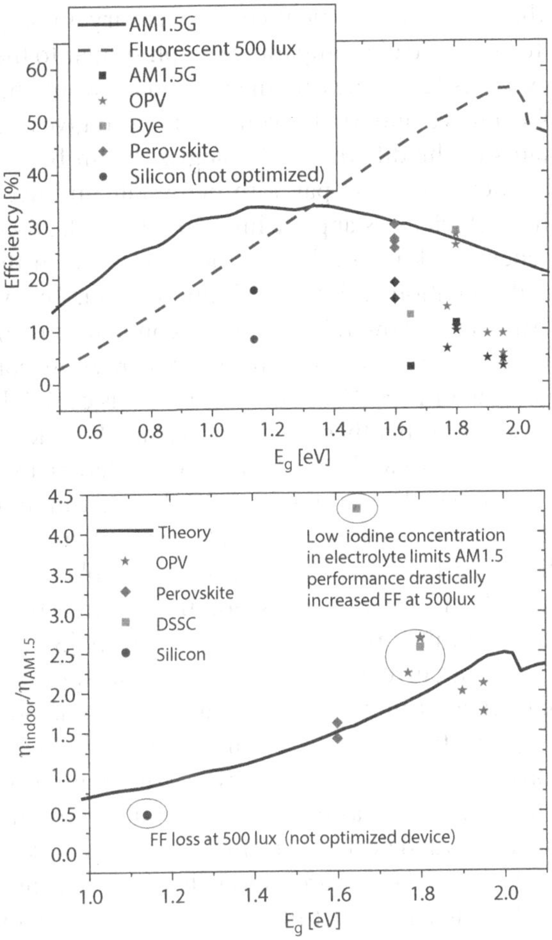

Figure 7.2 Experimental efficiencies of OPV, Dye and Perovskite PV cells taken from the literature as cited in the text compared to theoretical limits under AM1.5G (black solid line, calculated by the authors) and a fluorescent lamp (blue solid line, calculated by the authors). The different technologies are indicated by symbol shape and color for indoor measurements and only symbol shape (black) for their AM 1.5G efficiencies from the respective publications. It is important to note that the exact lighting conditions vary between different publications, such that this graph can only give a rough impression on the current status of the different technologies. Calculating the ratio between the indoor and outdoor efficiencies of the same device gives a consistency check of the low light measurements, as deviations from the theoretical curve need a corresponding reason (e.g., FF changes).

Although the exact measurement conditions vary among the different publications, to create a truly comparative graph similar to the well-known solar efficiency tables [27] a measurement standard would be required (as discussed in Chapters 3 and 5), this compilation can give a rough impression of the status of the different technologies. It can be seen that state-of-the-art OPV devices are on par with perovskite and dye photovoltaic cells, all reaching efficiencies approaching 30% [19, 20, 28–33]. An interesting consistency check is to calculate the ratio between the efficiencies under indoor illumination and AM1.5G illumination. This value is lower than one for bandgaps below 1.3 eV (e.g., Silicon, blue circle) and reaches around a factor of 2 at the optimum bandgap. Perovskite solar cells typically have a bandgap of 1.55 eV and the factor here is ca. 1.5. Experimental data lines up well with this theoretical curve, largely independent of the actual efficiency values (well reflected in the different low- and high-efficiency OPV devices). Deviations for OPV devices can in most cases be explained by significant changes in the fill factor (FF) due to mobility limitations, as discussed below; or in the case of dye photovoltaic cells, by a highly diluted electrolyte with a low iodine concentration [30], which leads to a diffusion limited maximum current density [34] which is lower than the short-circuit current under full sun illumination but larger than the short-circuit current under artificial light. If there is no obvious reason for a strong deviation, the measurements should be rechecked. Other possible reasons for a strong deviation would be nonlinear dependence of J_{SC} with light intensity, parasitic series or parallel resistance (R_S, R_P) or mechanisms which lead to deviation from the logarithmic V_{OC} dependence on light intensity. These effects are especially present in novel technologies with complex internal structures, such as OPV, Dye and Perovskite PV, while they are pretty much absent in, for example, Silicon and III-V semiconductor cells with close to perfect crystals and high charge carrier mobilities (see Chapter 9).

The second important implication of indoor lighting is that the light intensity is 2–3 orders of magnitude lower than for AM1.5G irradiation (1000 W/m^2). This is translated into a short circuit current density (J_{SC}) reduced by more or less the same factor. While under full sun illumination, current densities lie in the range of 10–40 mA/cm^2; under typical indoor lighting, the short-circuit current density is limited to only ca. 50–100 μA/cm^2 at 500 lux. Details depend on the exact light source and photovoltaic cell, as discussed in detail in the previous Chapters 3–5. These very low current densities have several implications. First of all, only very low consumption (μW–mW range) applications can be driven with reasonably sized photovoltaic cells (Chapter 2). Ohms law teaches two further important aspects. The photovoltaic cells must have a very high parallel resistance R_p, or else the parasitic current through this resistance easily reaches similar values as the generated current, leading to very low FF, reduced V_{OC} and thus low efficiency [26] (more details are discussed in Chapter 2). This issue is especially severe for OPV, as the soft polymeric layers are typically only 70–300 nm thin, and thus scratching is a concern in laboratory processes as well as in industrial production environments. Furthermore, the liquid coating process is prone to defects by dewetting, dust particles and roughness of underlying layers.

Figure 7.3 shows theoretical calculations of the maximum efficiency for a photovoltaic cell with a bandgap of 1.78 eV as a function of R_p. While under 1 sun illumination 10k Ωcm^2 is sufficient, for indoor light conditions this value would lead to nearly zero efficiency. Rp must be in the $M\Omega cm^2$ range to allow for high efficiencies, i.e., roughly a factor of 100 higher. This translates into the typical situation that the consistency of results is much higher at AM1.5G conditions than under low light measurements, where the often observed variations in R_p between 10^4–10^6 Wcm^2 result in strong variations in the efficiencies. Therefore, it is very important to find a robust material stack and processing which allows consistently achieving R_p values above 10^6 Wcm^2 by optimizing the coating quality or other strategies. To minimize the impact of shunts caused by coating and structuring defects, different strategies have been explored. A post treatment after coating of the semiconductor layer which locally reduces the conductivity of the underlying electrode where exposed was shown [35], and the use of two interlayers with low conductivity and/or high contact resistance [12, 36] has been employed in the modules shown in Figure 7.7.

A positive effect is that due to the low currents the resistance of the current extracting circuitry can be a factor of 100 higher than for solar applications, leading to strongly relaxed constraints for the dimensioning of cells and modules [16] and allowing the use of low conductivity materials (see Section 7.2).

A less obvious but related effect is that the charge carrier mobilities of the absorber materials can be quite low without compromising efficiency, as the low current can be carried by a lower bulk conductivity without significant losses. Typical OPV materials have low charge carrier mobilities in the range of 10^{-5}–10^{-3} cm^2/Vs, which indeed limit the fill factors and usable film thickness of these devices under full sunlight. More and more materials with higher charge carrier mobilities exceeding 1 cm^2/Vs have been found, allowing device thickness above 300 nm [37, 38]; however,

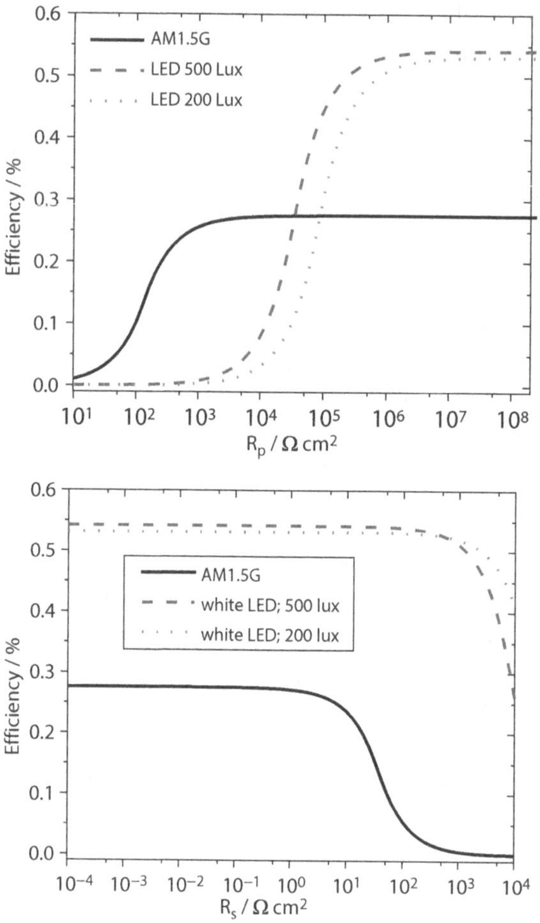

Figure 7.3 Efficiency of photovoltaic cells as a function of Rp for different illumination conditions. Under low light intensities, much higher Rp values are needed to maintain a high efficiency. In contrast, the series resistance can be substantially larger under indoor lighting, relaxing requirements for electrode conductivity and module dimensioning.

an increase in J_{SC} is still traded for lower FF, resulting in efficiency plateaus and not increasing efficiencies for thicker layers. The reason for the reduced FF is that for a given voltage V with $0 < V < V_{OC}$ (i.e., where the photovoltaic cell delivers power), a device with a low mobility absorber will accumulate more charge carriers. This leads to a quasi-Fermi level splitting in the bulk which is larger than the external voltage, resulting in enhanced recombination at this working point. This is reflected in a lower fill factor for such devices. Poor charge carrier transport can have a significant impact at high light intensity, but is much less an issue for typical indoor lighting conditions, thus broadening the range of suitable organic absorber materials and active layer thickness. Indeed, for materials with low charge carrier mobility often a significant increase of the fill factor, and hence power conversion efficiency, is observed for decreasing light intensity (see Figure 7.2). The impact of charge carrier mobility on device efficiency is shown in Figure 7.4 for the AM1.5G spectrum (left) as well as for a 500 lux white LED (corresponding to 1.62 W/m^2, right graph).

The absorber thickness in the simulation was set to 300 nm, see graphs which is a typical value for optimized devices. It can be seen that for

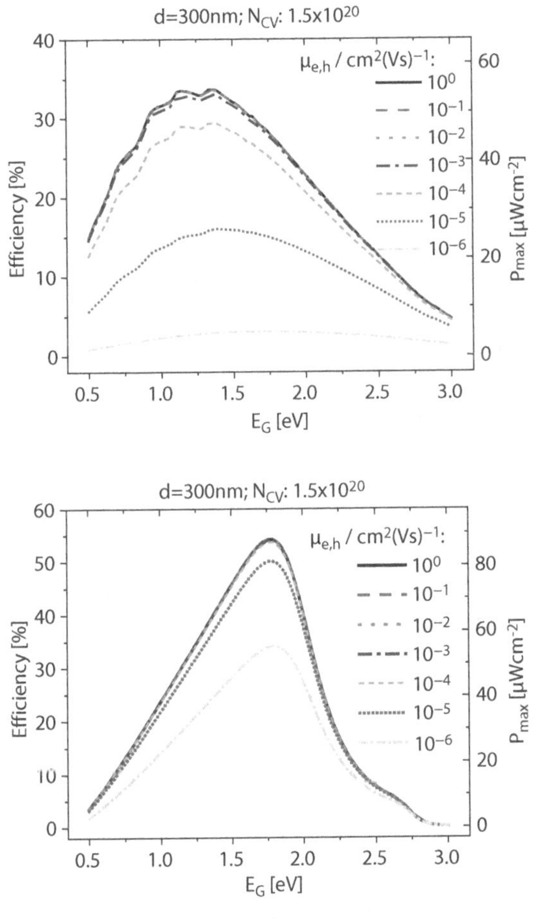

Figure 7.4 Maximum theoretical efficiency as function of bandgap for different charge carrier mobilities and for AM1.5G (top) as well as for a 500 lux white LED (bottom).

the AM1.5G spectrum the mobility should be 10^{-3} $cm^2(Vs)^{-1}$ or larger to avoid losses, whereas for the 500 lux white LED spectrum, mobilities $>10^{-5}$ $cm^2(Vs)^{-1}$ are sufficient. The calculations are based on an analytical approach presented in [39]. Note that the thicker the active layer, the larger the impact of the mobilities.

7.3 OPV Modules

For classical photovoltaic applications, very large area arrays (m^2) are built and extracting the generated electrical energy without significant ohmic losses is a challenge. The reason is that the current is proportional to the area of the device and the ohmic power loss is $P=R*I^2$. The solution is a combination of a highly conductive current collection scheme on cell level and series circuitry of single wafers (for c-Si) or cell strips (for thin films). Figure 7.5 shows this approach, as it is usually used for OPV modules as well.

For typical indoor applications the target areas are usually much smaller, in the range of cm^2, as the photovoltaic array should be integrated into a device in many cases. Furthermore, the low current densities relax the challenge for current extraction significantly, allowing the use of less conductive material and very large single cell units, leaving a series circuitry optional. This leads to a shift in design paradigm as described in the following.

For classical applications the task is to determine the optimum dimensions in order to achieve the best compromise between efficient energy extraction and shading by metal fingers and series circuitry, i.e., minimizing parasitic series resistance and area loss. In contrast, the design of integrated low light PV follows a different path; the focus lies on achieving the highest possible parallel resistance and meeting the system voltage (typically ~3.7V) and energy requirements of the electronics used. The available

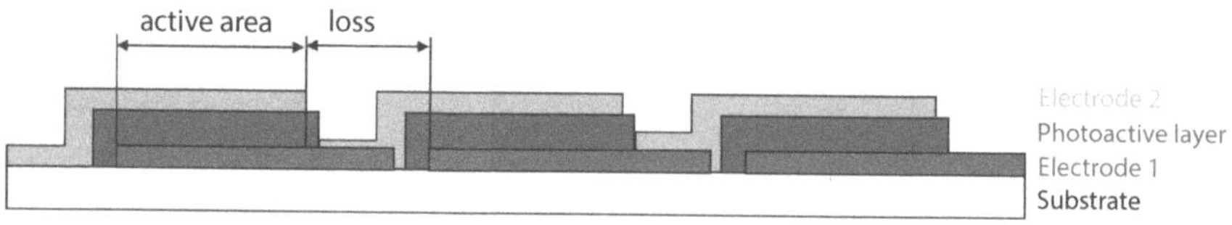

Figure 7.5 Monolithic series circuitry in a module. The photovoltaic array is patterned into cell strips, where the plus and minus pole of neighboring strips are connected to form a series circuitry. In this way the current is limited to the current of one strip, while the voltage multiplies with the number of cell strips.

area will be covered with a fixed number of cell strips, and the individual cells will be increased in area to generate more current, even if that will drive the dimensions out of the optimum.

The first question to answer is which area is needed to generate the necessary power and energy to reliably drive the functionality and whether this area is available. An important aspect is whether storage is employed or not, and which level of operational readiness is needed. If a high degree of operational readiness is desired without energy storage, the layout must be suited for functionality under worst case conditions. Using storage will shift the layout more to safely provide the needed power averaged over the typical conditions. Also, in this case it is of high importance to identify the longest possible period of time where the energy drawn from the battery exceeds the energy provided by the photovoltaic cell; photovoltaic array and battery size can be traded to some extent in this case. A pragmatic approach can be to cover as much area of the device as it is easily possible, because the cost of the integrated OPV will probably not be dominated by the area of it, but more by the system integration. Then the energy storage can be dimensioned to meet the application specific needs.

The second question is what voltage is desired in order to determine the number of series circuitries. It can of course only be adjusted in multiples of the unit cell voltage, as each cell strip in the module will add the same voltage. Taking the step from a single cell to a module leads to an area loss, where the films are patterned to overlap the plus and minus pole of neighboring cell strips (see Figure 7.5). This can be minimized by laser processing to less than 100 μm [40], but in typical printing and roll-to-roll production environments with limited registration accuracy and limited dimensional stability of the substrates it can easily exceed 1 mm. While this is a major concern for power applications, in the case of indoor PV this loss can be balanced by large cell areas. Another strategy is to employ only local series connections such that the cell strips are just separated by one line and the series circuitry is located only at the edge of the module [41].

The cell and module efficiency can be calculated as a function of sheet resistance of the electrodes and length of the cell strips using a numerical integration of the local diode currents and voltage losses due to the current flow [15, 16]. Typically, the transparent electrode is the limiting component, which is often formed by a transparent conductive oxide (TCO), a thin silver layer or polymeric conductive materials. Figure 7.6 shows the result of such calculations for single cells (black curves, squares) for a sheet resistance of 10 Ohm (solid line and symbol, typical for high quality TCOs and thin silver-based electrodes) and 1000 Ohm (dashed line and open symbol, easily reachable by solution processed TCOs and organic

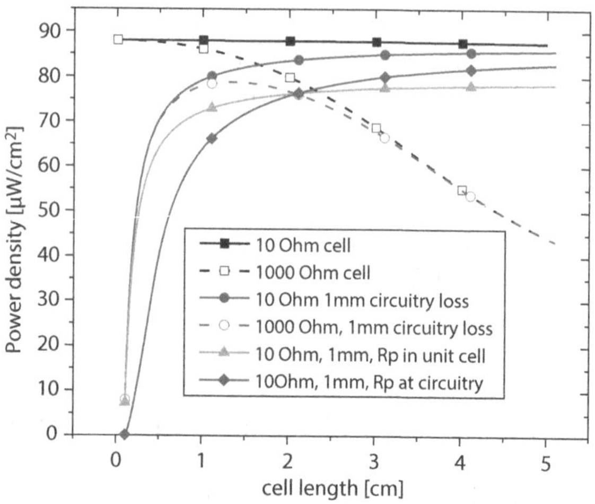

Figure 7.6 Simulation of device efficiency for single cells (black) and modules with 1 mm area loss (other colors). The input JV curve is the theoretical optimum solar cell. For real devices the maximum output power will be proportionally lower, but the trends will be the same.

transparent conductors). The red curves (circles) represent the same unit cell but integrated into a module with 1 mm loss for the series circuitry. Even for an electrode with 1000 Ohm sheet resistance a high fraction of the efficiency can be maintained. And with 10 Ohm one can practically reach the cell efficiency when the cell strips are more than 3 cm wide.

A low R_p in the unit cell globally reduces the obtainable efficiency (cyan, triangle), while a local shunt at the series circuitry can be partially compensated by larger device area (magenta, diamond).

The relaxed constraints for current extraction offer design freedom for shaping the cells to match the shape of the housing or just achieve a pleasant design, as shown in Figure 7.7. The possibility to coat and print the organic photovoltaic cells [42] offers unique flexibility in designing the shape of the photovoltaic cell—the production process is very similar to a label printing process—just using highly specialized materials and adapted machinery. Different coating and printing methods have been demonstrated ranging from slot-die [9, 21, 38, 43–46] and spray coating [47–50] to gravure [23, 51, 52] and inkjet printing [25, 53, 54]. Often printed silver (grid) electrodes are used processed with screen [55–56], flexographic [57, 58] or inkjet [53, 59] or aerosol [60] printing. For indoor photovoltaics this silver grid can be omitted, as the conductivity of the organic transparent conductor which is used in most cases in combination with the silver grid is conductive enough to collect the low currents over significant distances

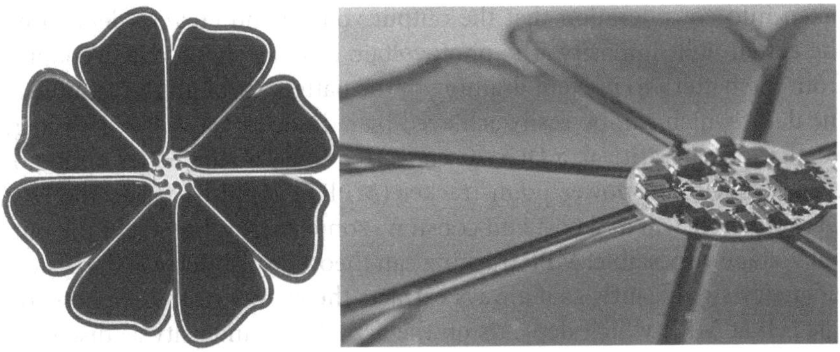

Figure 7.7 OPV flower designed by Fraunhofer ISE and gravure printed by VTT (left, © VTT) and equipped with a radio transmitting sensor by Wibicom (right, © Wibicom) during the EU-project ArtESun.

[16], especially if optimized formulations [61] and additives [62] or post treatments are used [63]. State-of-the-art materials reach values of typically a few hundred Ohm such that without problems single cells can be several centimeters wide before a current collecting feature, like a conductive metal track or a series circuitry, is needed.

7.4 OPV Devices and Applications

A large number of applications can be addressed with highly efficient, easy to integrate and pleasantly designable organic photovoltaic cells. The selling point would be the substitution of cables or batteries which need high upfront costs or costly maintenance. Photovoltaic cells can ideally be used to power devices which only need power when light is available, such that no other power source would be needed. Examples are pocket calculators, signage and sensors in supermarkets and stores, toilets and water taps; or sensors which have very low data frequencies (a few times per day, when gaps in the dark do not matter). A supercapacitor can be added when power peaks exceed the average/worst case power of the photovoltaic module. In more complex and long running applications the photovoltaic cell can significantly reduce the needed battery size or replacement intervals, as discussed in more detail in the proceeding Chapters 2 and 3. An example of a product evaluation is given in Chapter 8.

The electrical integration into devices can be achieved by direct wiring if the module is dimensioned to match the system voltage under typical light conditions. The direct wiring is simple and cheap, but it must be taken into consideration that the output voltage and current depend on the actual light intensity. The photovoltaic module should be decoupled from the battery to prevent draining of the battery through the module in the dark, which can be easily achieved by a diode or transistor. If voltage fluctuations are critical, a DC/DC converter could be the better option or even a maximum power point tracker (MPPT) could be used. The DC/DC converter will add cost and consume some energy, but keeps the output voltage very stable, MPP tracking can theoretically increase the system efficiency significantly as it always extracts the energy from the photovoltaic cell at MPP, which depends on the actual light intensity as discussed before. Nevertheless, for many low light applications the self-consumption of state-of-the-art converters or even MPPTs is prohibitive. The optimum dimensioning and layout are therefore application specific and highly dependent on the exact lighting conditions (i.e., the generation profile), the consumption profile, the degree of operational readiness required and the geometrical constraints of the device.

Mechanical integration can be achieved by adhesive film lamination [64] or, as in the example shown in Figure 7.7, the OPV device can carry the electronics.

A remark should be made with respect to the long-term stability of these devices. Although promising examples have been shown [65–75], the stability of these novel devices is highly dependent on the exact materials and processes used and thus needs verification for any new material and process. For the specific case of indoor photovoltaics only very few examples of long-term stability investigations are available [76], but it is highly probable that devices will last significantly longer than under outdoor exposure due to much milder environment in terms of exposure to light and UV light, moisture and temperatures.

7.5 Acceptance and Safety Considerations

Photovoltaic cells used in indoor and especially IoT applications are subject to more strict boundary conditions in terms of product safety and release into the environment than solar power plants. It is highly probably that IoT products will have to be compliant with RoHS regulations and it should always be considered that there is a much higher probability for these kinds of products to be disposed of incorrectly or even taken apart by people and in worst case by children. They should therefore not contain substances with acute toxicity and bioaccumulation potential. Typical OPV, Dye and Perovskite photovoltaic cells will contain only very little semiconductor material (on a mg-g scale), and from a purely scientific point of view one may come to the conclusion that the risk of an incident with severe consequences is very low. Nevertheless, from an acceptance point of view these considerations suggest that the use of Perovskite photovoltaic cells in the present form with a water-soluble and thus highly bioavailable lead component is unlikely to be accepted in such applications. The use of GaAs (apart from high cost) may as well be debated [77], although it is widely accepted in LEDs and other electronic devices, mainly due to the lack of alternatives. For DSSC and OPV it is more plausible that largely uncritical products can be designed [78–80] but it is by no means a granted built-in property. Aromatic organic compounds and liquid electrolytes can be toxic and may turn out to have a potentially negative impact in the long term as well. OPV seems to offer a high potential for uncritical products, as the organic semiconducting materials are solid and largely insoluble in water [79]. Nevertheless, this aspect should be taken into consideration during the development of the different novel technologies targeting consumer applications.

Another aspect are end-of-life scenarios. Recycling of OPV has been shown, but the environmental impact of such processes must be compared to the impact of just burning it in a thermal waste treatment, ideally in a combined heat and power unit to figure out which is the most eco-friendly choice [81]. As for indoor PV the use of silver can be omitted, it is highly probable that the disposal of the photovoltaic cells will be established as no valuable compound can be recovered. The substrate and encapsulating film will be 99% of the photovoltaic cell, which for indoor PV is probably PET. In principle this can be recycled well, but the photovoltaic cell will contain several other layers, namely the semiconducting polymers, metals and metal oxides as well as adhesives. To which extent these are problematic in a PET-recycling process will have to be investigated. Especially for indoor applications the use of nanopaper [82], bioplastics and even compostable films [83] can be an option as well. Compared to packaging films the total volumes will be marginal in any case.

References

1. Akamatu, H. and Inokuchi, H., On the electrical conductivity of violanthrone, iso-violanthrone, and pyranthrone. *J. Chem. Phys.*, 18, 6, 810, 1950. https://doi.org/10.1063/1.1747780.
2. Akamatu, H. and Inokuchi, H., Photoconductivity of violanthrone. *J. Chem. Phys.*, 20, 9, 1481–1483, 1952. https://doi.org/10.1063/1.1700784.

3. Akamatu, H., Inokuchi, H., Matsunaga, Y., Electrical conductivity of the perylene-bromine complex. *Nature (London, U.K.)*, 173, 4395, 168–169, 1954. https://doi.org/10.1038/173168a0.
4. Heeger, A.J., Nobel lecture: Semiconducting and metallic polymers: The fourth generation of polymeric materials. *Rev. Mod. Phys.*, 73, 3, 681–700, 2001. https://doi.org/10.1103/RevModPhys.73.681.
5. Kearns, D. and Calvin, M., Photovoltaic effect and photoconductivity in laminated organic systems. *J. Chem. Phys.*, 29, 4, 950–951, 1958. https://doi.org/10.1063/1.1744619.
6. Tang, C.W., Two-layer organic photovoltaic cell. *Appl. Phys. Lett.*, 48, 2, 183, 1986. https://doi.org/10.1063/1.96937.
7. Sariciftci, N.S., Smilowitz, L., Heeger, A.J., Wudl, F., Photoinduced electron transfer from a conducting polymer to buckminsterfullerene. *Science*, 258, 5087, 1474–1476, 1992. https://doi.org/10.1126/science.258.5087.1474.
8. Yu, G., Gao, J., Hummelen, J.C., Wudl, F., Heeger, A.J., Polymer photovoltaic cells: Enhanced efficiencies *via* a network of internal donor-acceptor heterojunctions. *Science*, 270, 5243, 1789–1791, 1995. https://doi.org/10.1126/science.270.5243.1789.
9. Zimmermann, B., Schleiermacher, H.-F., Niggemann, M., Würfel, U., ITO-free flexible inverted organic solar cell modules with high fill factor prepared by slot die coating. *Sol. Energy Mater. Sol. Cells*, 95, 7, 1587–1589, 2011. https://doi.org/10.1016/j.solmat.2010.11.025.
10. Galagan, Y., Rubingh, J.-E. J. M., Andriessen, R., Fan, C.-C., Blom, P.W.M., Veenstra, S.C., Kroon, J.M., ITO-free flexible organic solar cells with printed current collecting grids. *Sol. Energy Mater. Sol. Cells*, 95, 5, 1339–1343, 2011. https://doi.org/10.1016/j.solmat.2010.08.011.
11. Hösel, M., Angmo, D., Søndergaard, R.R., dos Reis Benatto, G.A., Carlé, J.E., Jørgensen, M., Krebs, F.C., High-volume processed, ITO-free superstrates and substrates for roll-to-roll development of organic electronics. *Adv. Sci.*, 1, 1400002, 2014. n/a https://doi.org/10.1002/advs.201400002.
12. Välimäki, M., Apilo, P., Po, R., Jansson, E., Bernardi, A., Ylikunnari, M., Vilkman, M., Corso, G., Puustinen, J., Tuominen, J., Hast, J., R2R-printed inverted OPV modules – towards arbitrary patterned designs. *Nanoscale*, 7, 21, 9570–9580, 2015. https://doi.org/10.1039/c5nr00204d.
13. NREL, Best Research-Cell Efficiency Chart. https://www.nrel.gov/pv/cell-efficiency.html (accessed 12 December 2019).
14. Vezie, M.S., Few, S., Meager, I., Pieridou, G., Dörling, B., Ashraf, R.S., Goñi, A.R., Bronstein, H., McCulloch, I., Hayes, S.C., Campoy-Quiles, M., Nelson, J., Exploring the origin of high optical absorption in conjugated polymers. *Nat. Mater.*, 15, 746–753, 2016. https://doi.org/10.1038/nmat4645.
15. Lungenschmied, C., Dennler, G., Neugebauer, H., Sariciftci, S.N., Glatthaar, M., Meyer, T., Meyer, A., Flexible, long-lived, large-area, organic solar cells. *Sol. Energy Mater. Sol. Cells*, 91, 5, 379–384, 2007. https://doi.org/10.1016/j.solmat.2006.10.013.

16. Niggemann, M., Zimmermann, B., Haschke, J., Glatthaar, M., Gombert, A., Organic solar cell modules for specific applications—From energy autonomous systems to large area photovoltaics. *Thin Solid Films*, 516, 20, 7181–7187, 2008. https://doi.org/10.1016/j.tsf.2007.12.093.
17. Freunek, M., Freunek, M., Reindl, L.M., Maximum efficiencies of indoor photovoltaic devices. *IEEE J. Photovoltaics*, 3, 1, 59–64, 2013. https://doi.org/10.1109/JPHOTOV.2012.2225023.
18. Azzouzi, M., Kirchartz, T., Nelson, J., Factors controlling open-circuit voltage losses in organic solar cells. *Trends Chem.*, 1, 1, 49–62, 2019. https://doi.org/10.1016/j.trechm.2019.01.010.
19. Cui, Y., Wang, Y., Bergqvist, J., Yao, H., Xu, Y., Gao, B., Yang, C., Zhang, S., Inganäs, O., Gao, F., Hou, J., Wide-gap non-fullerene acceptor enabling high-performance organic photovoltaic cells for indoor applications. *Nat. Energy*, 4, 9, 768–775, 2019. https://doi.org/10.1038/s41560-019-0448-5.
20. Lee, H.K.H., Wu, J., Barbé, J., Jain, S.M., Wood, S., Speller, E.M., Li, Z., Castro, F.A., Durrant, J.R., Tsoi, W.C., Organic photovoltaic cells – promising indoor light harvesters for self-sustainable electronics. *J. Mater. Chem. A*, 6, 14, 5618–5626, 2018. https://doi.org/10.1039/C7TA10875C.
21. Krebs, F.C., Polymer solar cell modules prepared using roll-to-roll methods: Knife-over-edge coating, slot-die coating and screen printing. *Sol. Energy Mater. Sol. Cells*, 93, 4, 465–475, 2009. https://doi.org/10.1016/j.solmat.2008.12.012.
22. Krebs, F.C., Spanggard, H., Kjær, T., Biancardo, M., Alstrup, J., Large area plastic solar cell modules. *Mater. Sci. Eng., B*, 138, 2, 106–111, 2007. https://doi.org/10.1016/j.mseb.2006.06.008.
23. Kopola, P., Aernouts, T., Guillerez, S., Jin, H., Tuomikoski, M., Maaninen, A., Hast, J., High efficient plastic solar cells fabricated with a high-throughput gravure printing method. *Org. Electron.*, 94, 10, 1673–1680, 2010. https://doi.org/10.1016/j.solmat.2010.05.027.
24. Galagan, Y., de Vries, I.G., Langen, A.P., Andriessen, R., Verhees, W.J.H., Veenstra, S.C., Kroon, J.M., Technology development for roll-to-roll production of organic photovoltaics. *Chem. Eng. Process*, 50, 5-6, 454–461, 2011. https://doi.org/10.1016/j.cep.2010.07.012.
25. Eggenhuisen, T.M., Galagan, Y., Biezemans, A.F.K.V., Slaats, T.M.W.L., Voorthuijzen, W.P., Kommeren, S., Shanmugam, S., Teunissen, J.P., Hadipour, A., Verhees, W.J.H., Veenstra, S.C., Coenen, M.J.J., Gilot, J., Andriessen, R., Groen, W.A., High efficiency, fully inkjet printed organic solar cells with freedom of design. *J. Mater. Chem. A*, 3, 14, 7255–7262, 2015. https://doi.org/10.1039/c5ta00540j.
26. Steim, R., Ameri, T., Schilinsky, P., Waldauf, C., Dennler, G., Scharber, M., Brabec, C.J., Organic photovoltaics for low light applications. *Sol. Energy Mater. Sol. Cells*, 95, 12, 3256–3261, 2011. https://doi.org/10.1016/j.solmat.2011.07.011.
27. Green, M.A., Dunlop, E.D., Levi, D.H., Hohl-Ebinger, J., Yoshita, M., Ho-Baillie, A.W.Y., Solar cell efficiency tables (version 54). *Prog. Photovolt. Res. Appl.*, 27, 7, 565–575, 2019. https://doi.org/10.1002/pip.3171.

28. Dagar, J., Castro-Hermosa, S., Lucarelli, G., Cacialli, F., Brown, T.M., Highly efficient perovskite solar cells for light harvesting under indoor illumination *via* solution processed SnO2/MgO composite electron transport layers. *Nano Energy*, 49, 290–299, 2018. https://doi.org/10.1016/j.nanoen.2018.04.027.
29. Freitag, M., Teuscher, J., Saygili, Y., Zhang, X., Giordano, F., Liska, P., Hua, J., Zakeeruddin, S.M., Moser, J.-E., Grätzel, M., Hagfeldt, A., Dye-sensitized solar cells for efficient power generation under ambient lighting. *Nature Photon*, 11, 6, 372–378, 2017. https://doi.org/10.1038/nphoton.2017.60.
30. de Rossi, F., Pontecorvo, T., Brown, T.M., Characterization of photovoltaic devices for indoor light harvesting and customization of flexible dye solar cells to deliver superior efficiency under artificial lighting. *Appl. Energy*, 156, 413–422, 2015. https://doi.org/10.1016/j.apenergy.2015.07.031.
31. Chen, C.-Y., Lee, W.-H., Hsiao, S.-Y., Tsai, W.-L., Yang, L., Lin, H.-L., Chou, H.-J., Lin, H.-W., Vacuum-deposited perovskite photovoltaics for highly efficient environmental light energy harvesting. *J. Mater. Chem. A*, 7, 8, 3612–3617, 2019. https://doi.org/10.1039/C8TA11515J.
32. Chen, C.-Y., Chang, J.-H., Chiang, K.-M., Lin, H.-L., Hsiao, S.-Y., Lin, H.-W., Perovskite photovoltaics for dim-light applications. *Adv. Funct. Mater.*, 25, 45, 7064–7070, 2015. https://doi.org/10.1002/adfm.201503448.
33. Di Giacomo, F., Zardetto, V., Lucarelli, G., Cinà, L., Di Carlo, A., Creatore, M., Brown, T.M., Mesoporous perovskite solar cells and the role of nanoscale compact layers for remarkable all-round high efficiency under both indoor and outdoor illumination. *Nano Energy*, 30, 460–469, 2016. https://doi.org/10.1016/j.nanoen.2016.10.030.
34. Hinsch, A., Behrens, S., Berginc, M., Bönnemann, H., Brandt, H., Drewitz, A., Einsele, F., Faßler, D., Gerhard, D., Gores, H., Haag, R., Herzig, T., Himmler, S., Khelashvili, G., Koch, D., Nazmutdinova, G., Opara-Krasovec, U., Putyra, P., Rau, U., Sastrawan, R., Schauer, T., Schreiner, C., Sensfuss, S., Siegers, C., Skupien, K., Wachter, P., Walter, J., Wasserscheid, P., Würfel, U., Zistler, M., Material development for dye solar modules: Results from an integrated approach. *Prog. Photovolt. Res. Appl.*, 16, 6, 489–501, 2008. https://doi.org/10.1002/pip.832.
35. Jolt Oostra, A., Blom, P.W.M., Michels, J.J., Prevention of short circuits in solution-processed OLED devices. *Organic Electronics*, 15, 6, 1166–1172, 2014. https://doi.org/10.1016/j.orgel.2014.03.008.
36. Apilo, P., Hiltunen, J., Välimäki, M., Heinilehto, S., Sliz, R., Hast, J., Roll-to-roll gravure printing of organic photovoltaic modules-insulation of processing defects by an interfacial layer. *Prog. Photovolt: Res. Appl.*, 23, 918–928, 2015, 2014. n/ahttps://doi.org/10.1002/pip.2508.
37. Chen, Z., Cai, P., Chen, J., Liu, X., Zhang, L., Lan, L., Peng, J., Ma, Y., Cao, Y., Low band-gap conjugated polymers with strong interchain aggregation and very high hole mobility towards highly efficient thick-film polymer solar cells. *Adv. Mater.*, 26, 16, 2586–2591, 2014. https://doi.org/10.1002/adma.201305092.

38. Berny, S., Blouin, N., Distler, A., Egelhaaf, H.-J., Krompiec, M., Lohr, A., Lozman, O.R., Morse, G.E., Nanson, L., Pron, A., Sauermann, T., Seidler, N., Tierney, S., Tiwana, P., Wagner, M., Wilson, H., Solar trees: First large-scale demonstration of fully solution coated, semitransparent, flexible organic photovoltaic modules. *Adv. Sci*, 3, 1500342, 2016, 2015. https://doi.org/10.1002/advs.201500342.
39. Würfel, U., Neher, D., Spies, A., Albrecht, S., Impact of charge transport on current-voltage characteristics and power-conversion efficiency of organic solar cells. *Nat. Commun.*, 6, 6951, 2015. https://doi.org/10.1038/ncomms7951.
40. Lucera, L., Machui, F., Kubis, P., Schmidt, H.D., Adams, J., Strohm, S., Ahmad, T., Forberich, K., Egelhaaf, H.-J., Brabec, C.J., Highly efficient, large area, roll coated flexible and rigid OPV modules with geometric fill factors up to 98.5% processed with commercially available materials. *Energy Environ. Sci.*, 9, 1, 89–94, 2016. https://doi.org/10.1039/C5EE03315B.
41. Aoki, Y., Photovoltaic performance of organic photovoltaics for indoor energy harvester. *Organic Electronics*, 48, 194–197, 2017. https://doi.org/10.1016/j.orgel.2017.05.023.
42. Krebs, F.C., Fabrication and processing of polymer solar cells: A review of printing and coating techniques. *Sol. Energy Mater. Sol. Cells*, 93, 4, 394–412, 2009. https://doi.org/10.1016/j.solmat.2008.10.004.
43. Krebs, F.C., Espinosa, N., Hösel, M., Søndergaard, R.R., Jørgensen, M., 25th Anniversary Article: Rise to power – OPV-based solar parks. *Adv. Mater.*, 26, 1, 29–39, 2014. https://doi.org/10.1002/adma.201302031.
44. Ciro, J., Mejía-Escobar, M.A., Jaramillo, F., Slot-die processing of flexible perovskite solar cells in ambient conditions. *Sol. Energy*, 150, 570–576, 2017. https://doi.org/10.1016/j.solener.2017.04.071.
45. Kaduwal, D., Schleiermacher, H.-F., Schulz-Gericke, J., Schiefer, S., Liang Tan, Y., Zhang, J., Zimmermann, B., Würfel, U., Layout flexibility for sheet-to-sheet produced flexible ITO-free organic solar modules with organic functional layers slot die coated under ambient atmospheric conditions. *Sol. Energy Mater. Sol. Cells*, 136, 200–205, 2015. https://doi.org/10.1016/j.solmat.2014.11.043.
46. Machui, F., Lucera, L., Spyropoulos, G.D., Cordero, J., Ali, A.S., Kubis, P., Ameri, T., Voigt, M.M., Brabec, C.J., Large area slot-die coated organic solar cells on flexible substrates with non-halogenated solution formulations. *Sol. Energy Mater. Sol. Cells*, 128, 441–446, 2014. https://doi.org/10.1016/j.solmat.2014.06.017.
47. Reale, A., La Notte, L., Salamandra, L., Polino, G., Susanna, G., Brown, T.M., Brunetti, F., Di Carlo, A., Spray coating for polymer solar cells: An up-to-date overview. *Energy Technol.*, 3, 4, 385–406, 2015. https://doi.org/10.1002/ente.201402180.
48. Green, R., Morfa, A., Ferguson, A.J., Kopidakis, N., Rumbles, G., Shaheen, S.E., Performance of bulk heterojunction photovoltaic devices prepared by

airbrush spray deposition. *Appl. Phys. Lett.*, 92, 3, 33301, 2008. https://doi.org/10.1063/1.2836267.

49. Bishop, J.E., Mohamad, D.K., Wong-Stringer, M., Smith, A., Lidzey, D.G., Spray-cast multilayer perovskite solar cells with an active-area of 1.5 cm2. *Sci. Rep.*, 7, 1, 7962, 2017. https://doi.org/10.1038/s41598-017-08642-2.
50. Susanna, G., Salamandra, L., Brown, T.M., Di Carlo, A., Brunetti, F., Reale, A., Airbrush spray-coating of polymer bulk-heterojunction solar cells. *Sol. Energy Mater. Sol. Cells*, 95, 7, 1775–1778, 2011. https://doi.org/10.1016/j.solmat.2011.01.047.
51. Kapnopoulos, C., Mekeridis, E.D., Tzounis, L., Polyzoidis, C., Zachariadis, A., Tsimikli, S., Gravalidis, C., Laskarakis, A., Vouroutzis, N., Logothetidis, S., Fully gravure printed organic photovoltaic modules: A straightforward process with a high potential for large scale production. *Sol. Energy Mater. Sol. Cells*, 144, 724–731, 2016. https://doi.org/10.1016/j.solmat.2015.10.021.
52. Noh, J., Yeom, D., Lim, C., Cha, H., Han, J., Kim, J., Park, Y., Subramanian, V., Cho, G., Scalability of roll-to-roll gravure-printed electrodes on plastic foils. *IEEE Trans. Electron. Packag. Manufact.*, 33, 4, 275–283, 2010. https://doi.org/10.1109/TEPM.2010.2057512.
53. Galagan, Y., Coenen, E.W.C., Sabik, S., Gorter, H.H., Barink, M., Veenstra, S.C., Kroon, J.M., Andriessen, R., Blom, P.W.M., Evaluation of ink-jet printed current collecting grids and busbars for ITO-free organic solar cells. *Sol. Energy Mater. Sol. Cells*, 104, 32–38, 2012. https://doi.org/10.1016/j.solmat.2012.04.039.
54. Gheno, A., Huang, Y., Bouclé, J., Ratier, B., Rolland, A., Even, J., Vedraine, S., Toward highly efficient inkjet-printed perovskite solar cells fully processed under ambient conditions and at low temperature. *Sol. RRL*, 2, 11, 1800191, 2018. https://doi.org/10.1002/solr.201800191.
55. Krebs, F.C., Jørgensen, M., Norrman, K., Hagemann, O., Alstrup, J., Nielsen, T.D., Fyenbo, J., Larsen, K., Kristensen, J., A complete process for production of flexible large area polymer solar cells entirely using screen printing—First public demonstration. *Sol. Energy Mater. Sol. Cells*, 93, 4, 422–441, 2009. https://doi.org/10.1016/j.solmat.2008.12.001.
56. Kim, J., Duraisamy, N., Lee, T.-M., Kim, I., Choi, K.-H., Screen printed silver top electrode for efficient inverted organic solar cells. *Mater. Res. Bull.*, 70, 412–415, 2015. https://doi.org/10.1016/j.materresbull.2015.04.052.
57. Hösel, M., Søndergaard, R.R., Angmo, D., Krebs, F.C., Comparison of fast roll-to-roll flexographic, inkjet, flatbed, and rotary screen printing of metal back electrodes for polymer solar cells. *Adv. Eng. Mater.*, 15, 10, 995–1001, 2013. https://doi.org/10.1002/adem.201300011.
58. Krebs, F.C., Fyenbo, J., Jørgensen, M., Product integration of compact roll-to-roll processed polymer solar cell modules: Methods and manufacture using flexographic printing, slot-die coating and rotary screen printing. *J. Mater. Chem.*, 20, 41, 8994, 2010. https://doi.org/10.1039/c0jm01178a.

59. Yu, J.-S., Kim, I., Kim, J.-S., Jo, J., Larsen-Olsen, T.T., Søndergaard, R.R., Hösel, M., Angmo, D., Jørgensen, M., Krebs, F.C., Silver front electrode grids for ITO-free all printed polymer solar cells with embedded and raised topographies, prepared by thermal imprint, flexographic and inkjet roll-to-roll processes. *Nanoscale*, 4, 19, 6032–6040, 2012. https://doi.org/10.1039/c2nr31508d.
60. Kopola, P., Zimmermann, B., Filipovic, A., Schleiermacher, H.-F., Greulich, J., Rousu, S., Hast, J., Myllylä, R., Würfel, U., Aerosol jet printed grid for ITO-free inverted organic solar cells. *Sol. Energy Mater. Sol. Cells*, 107, 252–258, 2012. https://doi.org/10.1016/j.solmat.2012.06.042.
61. Stöcker, T., Köhler, A., Moos, R., Why does the electrical conductivity in PEDOT : PSS decrease with PSS content? A study combining thermoelectric measurements with impedance spectroscopy. *J. Polym. Sci., Part B: Polym. Phys.*, 50, 14, 976–983, 2012. https://doi.org/10.1002/polb.23089.
62. Li, Z., Meng, W., Tong, J., Zhao, C., Qin, F., Jiang, F., Xiong, S., Zeng, S., Xu, L., Hu, B., Zhou, Y., A nonionic surfactant simultaneously enhancing wetting property and electrical conductivity of PEDOT : PSS for vacuum-free organic solar cells. *Sol. Energy Mater. Sol. Cells*, 137, 311–318, 2015. https://doi.org/10.1016/j.solmat.2015.02.024.
63. Dimitriev, O.P., Grinko, D.A., Noskov, Y.V., Ogurtsov, N.A., Pud, A.A., PEDOT : PSS films—Effect of organic solvent additives and annealing on the film conductivity. *Synth. Met*, 159, 21-22, 2237–2239, 2009. https://doi.org/10.1016/j.synthmet.2009.08.022.
64. Krebs, F.C., Fyenbo, J., Tanenbaum, D.M., Gevorgyan, S.A., Andriessen, R., van Remoortere, B., Galagan, Y., Jørgensen, M., The OE-A OPV demonstrator anno domini 2011. *Energy Environ. Sci.*, 4, 10, 4116, 2011. https://doi.org/10.1039/c1ee01891d.
65. Zimmermann, B., Würfel, U., Niggemann, M., Longterm stability of efficient inverted P3HT : PCBM solar cells. *Sol. Energy Mater. Sol. Cells*, 93, 4, 491–496, 2009. https://doi.org/10.1016/j.solmat.2008.12.022.
66. Sapkota, S.B., Spies, A., Zimmermann, B., Dürr, I., Würfel, U., Promising long-term stability of encapsulated ITO-free bulk-heterojunction organic solar cells under different aging conditions. *Sol. Energy Mater. Sol. Cells*, 130, 144–150, 2014. https://doi.org/10.1016/j.solmat.2014.07.004.
67. Fu, Q., Tang, X., Huang, B., Hu, T., Tan, L., Chen, L., Chen, Y., Recent progress on the long-term stability of perovskite solar cells. *Adv. Sci.*, 5, 5, 1700387, 2018. https://doi.org/10.1002/advs.201700387.
68. Chen, L., Tan, Y.-Y., Chen, Z.-X., Wang, T., Hu, S., Nan, Z.-A., Xie, L.-Q., Hui, Y., Huang, J.-X., Zhan, C., Wang, S.-H., Zhou, J.-Z., Yan, J.-W., Mao, B.-W., Tian, Z.-Q., Toward long-term stability: Single-crystal alloys of cesium-containing mixed cation and mixed halide perovskite. *J. Am. Chem. Soc.*, 141, 4, 1665–1671, 2019. https://doi.org/10.1021/jacs.8b11610.
69. Bai, S., Da, P., Li, C., Wang, Z., Yuan, Z., Fu, F., Kawecki, M., Liu, X., Sakai, N., Wang, J.T.-W., Huettner, S., Buecheler, S., Fahlman, M., Gao, F., Snaith,

H.J., Planar perovskite solar cells with long-term stability using ionic liquid additives. *Nature (London, U.K.)*, 571, 7764, 245–250, 2019. https://doi.org/10.1038/s41586-019-1357-2.

70. Mateker, W.R. and McGehee, M.D., Progress in understanding degradation mechanisms and improving stability in organic photovoltaics. *Adv. Mater.*, 29, 10, 1603940, 2017. https://doi.org/10.1002/adma.201603940.
71. Angmo, D., Gevorgyan, S.A., Larsen-Olsen, T.T., Søndergaard, R.R., Hösel, M., Jørgensen, M., Gupta, R., Kulkarni, G.U., Krebs, F.C., Scalability and stability of very thin, roll-to-roll processed, large area, indium-tin-oxide free polymer solar cell modules. *Org. Electron.*, 14, 3, 984–994, 2013. https://doi.org/10.1016/j.orgel.2012.12.033.
72. Angmo, D., Sommeling, P.M., Gupta, R., Hösel, M., Gevorgyan, S.A., Kroon, J.M., Kulkarni, G.U., Krebs, F.C., Outdoor operational stability of indium-free flexible polymer solar modules over 1 year studied in India, Holland, and Denmark. *Adv. Eng. Mater.*, 16, 8, 976–987, 2014. https://doi.org/10.1002/adem.201400002.
73. Peters, C.H., Sachs-Quintana, I.T., Kastrop, J.P., Beaupré, S., Leclerc, M., McGehee, M.D., High efficiency polymer solar cells with long operating lifetimes. *Adv. Energy Mater.*, 1, 4, 491–494, 2011. https://doi.org/10.1002/aenm.201100138.
74. Peters, C.H., Sachs-Quintana, I.T., Mateker, W.R., Heumueller, T., Rivnay, J., Noriega, R., Beiley, Z.M., Hoke, E.T., Salleo, A., McGehee, M.D., The mechanism of burn-in loss in a high efficiency polymer solar cell. *Adv. Mater.*, 24, 5, 663–668, 2012. https://doi.org/10.1002/adma.201103010.
75. Mateker, W.R., Douglas, J.D., Cabanetos, C., Sachs-Quintana, I.T., Bartelt, J.A., Hoke, E.T., El Labban, A., Beaujuge, P.M., Fréchet, J.M.J., McGehee, M.D., Improving the long-term stability of PBDTTPD polymer solar cells through material purification aimed at removing organic impurities. *Energy Environ. Sci.*, 6, 8, 2529, 2013. https://doi.org/10.1039/c3ee41328d.
76. Zhang, Y., Yi, H., Iraqi, A., Kingsley, J., Buckley, A., Wang, T., Lidzey, D.G., Comparative indoor and outdoor stability measurements of polymer based solar cells. *Sci. Rep.*, 7, 1, 1305, 2017. https://doi.org/10.1038/s41598-017-01505-w.
77. Tanaka, A., Toxicity of indium arsenide, gallium arsenide, and aluminium gallium arsenide. *Toxicol. Appl. Pharmacol.*, 198, 3, 405–411, 2004. https://doi.org/10.1016/j.taap.2003.10.019.
78. Espinosa, N., Zimmermann, Y.-S., dos Reis Benatto, G.A., Lenz, M., Krebs, F.C., Outdoor fate and environmental impact of polymer solar cells through leaching and emission to rainwater and soil. *Energy Environ. Sci.*, 9, 5, 1674–1680, 2016. https://doi.org/10.1039/C6EE00578K.
79. Zimmermann, Y.-S., Schaffer, A., Hugi, C., Fent, K., Corvini, P.F.-X., Lenz, M., Organic photovoltaics: Potential fate and effects in the environment. *Environ. Int.*, 49, 128–140, 2012. https://doi.org/10.1016/j.envint.2012.08.015.

80. Zimmermann, Y.-S., Schäffer, A., Corvini, P.F.-X., Lenz, M., Thin-film photovoltaic cells: Long-term metal (loid) leaching at their end-of-life. *Environ. Sci. Technol.*, 47, 22, 13151–13159, 2013. https://doi.org/10.1021/es402969c.
81. Søndergaard, R.R., Espinosa, N., Jørgensen, M., Krebs, F.C., Efficient decommissioning and recycling of polymer solar cells: Justification for use of silver. *Energy Environ. Sci.*, 7, 3, 1006, 2014. https://doi.org/10.1039/c3ee43746a.
82. Nogi, M., Karakawa, M., Komoda, N., Yagyu, H., Nge, T.T., Transparent conductive nanofiber paper for foldable solar cells. *Sci. Rep.*, 5, 17254, 2015. https://doi.org/10.1038/srep17254.
83. Biodegradable polymer solar cells. https://www.sciencedirect.com/science/article/pii/S0927024808000366?via%3Dihub (accessed 19 December 2019).

8

High-Efficiency Indoor Photovoltaic Energy Harvesting

Matthias Kauer* **and Mathieu Bellanger**

Lightricity Ltd, Oxford Science Park, Oxford, U.K.

Abstract

An indoor photovoltaic energy harvesting device technology has been developed that achieves high efficiencies and high power per area at low light levels. A 50 mm x 20 mm energy harvesting device comprising a compound semiconductor material is demonstrated. At 200 lux white LED illumination (CCT 2,831 K), its operating power at maximum power point is 170.0 μW with V_{oc} = 4.61 V and I_{sc} = 43.8 μA, corresponding to an active area power density of 21.8 μW/cm^2.

Based on this high-efficiency energy harvesting device, an indoor photovoltaic (PV) energy harvesting power module (IPEHPM) has been developed for autonomously powering IoT sensor nodes and other devices which ordinarily would require disposable or user-rechargeable batteries to operate. The IPEHPM can deliver 3.6 ~ 4.2 V output voltage, with 100 mA pulse current for up to 600 ms. An average energy harvester to storage efficiency of 88.7% has been demonstrated at 200 lux. We demonstrate that the IPEHPM is suitable for perpetually powering a wireless building automation CO_2 sensor node at 200 lux, with CO_2 measurement and data transmission every 150 s.

Applications of high-efficiency energy harvesting devices in wearables and in wireless beacons are also discussed.

Keywords: Photovoltaic (PV) energy harvesting, Internet of Things (IoT), energy storage, power management

**Corresponding author*: matthias.kauer@lightricity.co.uk

Monika Freunek Müller (ed.) Indoor Photovoltaics: Materials, Modeling and Applications, (213–240)

8.1 Introduction

Energy harvesting addresses the needs and challenges of the fast-growing Internet-of-Things market, particularly in wearables, smart sensing and monitoring, tracking and positioning of assets, building and home automation and retail applications. The main customer and industry pain point is the current need to either have a wired power connection, which is unpractical or even infeasible in some use cases, or to replace (or manually recharge) the batteries in IoT sensors, devices and products. Battery replacement involves labor, transport and chemical waste that have detrimental environmental impact, including increased CO_2 emissions and financial consequences via the increased cost-of-ownership due to additional maintenance requirements.

Energy harvesting is ultimately required for sustainably powering all use cases requiring sensing, information-storage or transmission at scale. For example, in consumer retail, use cases extend from the in-store environment via intelligent shelf-labels and retail displays into manufacturing and logistics, via logistics and supply sensors. Many applications, whether they are in commercial and residential buildings, factories, warehouses, hospitals or supermarkets are tied to an indoor environment and there is no current solution for harvesting efficiently ambient indoor light as the low-level light of typically 10–1000 lux is a challenge. Wearable devices, whether they are smart-watches, medical wearables, fitness trackers or occupational safety devices face a similar challenge, in that they are worn in indoor environments for most of the time they are used.

There is thus a large gap in the market for a compact, cost-effective, battery-free system that can operate efficiently and be easily and seamlessly retrofitted inside buildings of any kind, which indoor PV energy harvesting technology can target.

High-efficiency indoor PV energy harvesting could thus eliminate the need for primary battery replacements or manual recharging of secondary batteries, reducing operational, maintenance and environmental costs in large-scale deployments, saving millions of batteries per year.

Commercially available light energy harvesting devices to date have been almost exclusively based on Silicon technology, with amorphous (a-Si) or crystalline (c-Si) Silicon providing output powers of ~5 $\mu W/cm^2$ under typical indoor lighting conditions (200 lux).

But the performance of current light energy harvesting technology is insufficient to meet the power requirements of many current IoT devices, from smart-watches and wearables with sensing, wireless capability and high-resolution displays (>50 μW average power consumption), asset trackers with long-range wireless protocols (up to ~100 μW average power consumption) to building automation sensors with gas sensors and wireless connectivity (up to ~150 μW average power consumption).

More recently, light energy harvesting technologies with higher efficiency than silicon-based technologies are being developed. For example, Lightricity's energy harvesting technology converts very efficiently the ambient light into electricity even under artificial low-light levels experienced inside buildings and homes (50–500 lux), with >30% efficiency or up to 6x more power per area than conventional silicon-based technologies at 200 lux demonstrated.

Such a high-efficiency PV energy harvesting device can then be combined with ultra-low power electronics and storage to provide a plug-and-play power solution, i.e., everlasting source of power, that directly connects to an IoT sensor or device. As a result, for a product designed to be otherwise powered by a coin-cell battery, up to 80 battery disposals per product could be avoided over a 10-year period, saving hundreds of euros per device.

As was discussed in detail in Chapters 2 and 3, there are other methods of harvesting that could compete with PV harvesting, especially in industrial applications. These rely on either thermal gradients [1] (thermoelectric generators), vibrations [2] (electrodynamic, piezoelectric) or electromagnetic fields [3] (wireless RF energy). However, thermoelectric generators require a heatsink and are too bulky to be used in many applications. RF chargers and piezoelectric generators have typically smaller sizes that are suitable for integration, but have power densities at least one order of magnitude less than PV energy harvesting when they rely on a standard environment, or rely on an external high power source such as a high magnetic flux, or a mechanical excitation to deliver greater amounts of power.

Energy harvesting has been used for decades in solar panels or even longer in bicycle dynamos, but the field has seen a resurgence of interest and new technologies during recent years with applications ranging from industry automation, smart city applications and many other Internet of Things (IoT) applications [4]. The challenge in the field of light energy harvesting is to create a manufacturable energy harvesting solution that can provide the required combination of performance, small size and low cost (sub-dollar) for the billions of connected sensors that are widely forecast to be deployed in the coming years [5].

Most of the materials and device technologies used for outdoor solar power generation have been investigated for indoor energy harvesting, too. However, the far lower energy available in the light spectrum indoors compared with the outdoor solar spectrum and the substantially different light spectra found in indoor environments mean that an efficient outdoor solar technology is generally not the most efficient solution for indoor harvesting.

In this chapter, we will provide an overview of Lightricity's approach for high-efficiency indoor light harvesting. We will then discuss technical approaches for integrating a light energy harvesting system into a power sub-system, comprising energy harvesting device, power management and storage. We will also describe in detail an autonomous building automation sensor and will discuss performance of light energy harvesting in this application scenario.

8.2 Approaches for Efficient Indoor PV Energy Harvesting

8.2.1 PV Energy Harvesting Technologies

In this section we will review the most commonly used or researched materials for light energy harvesting devices.

Amorphous (a-Si) [6] and crystalline (c-Si) silicon provide output powers of 3–6 μW/cm^2 under typical indoor lighting conditions (200 lux), corresponding to an efficiency of 5–10% indoors. They typically present a blueish (c-Si) or brownish color (a-Si) caused by imperfect visible light absorption and surface reflections (Figure 8.1 (a)). Amorphous silicon (a-Si) is commercially available but its efficiency is reduced under very low-level indoor lighting (<200 lux) conditions, as a-Si suffers a "cut-out" at lower limit of illumination, when charge carrier mobility is low, and recombination is a major issue.

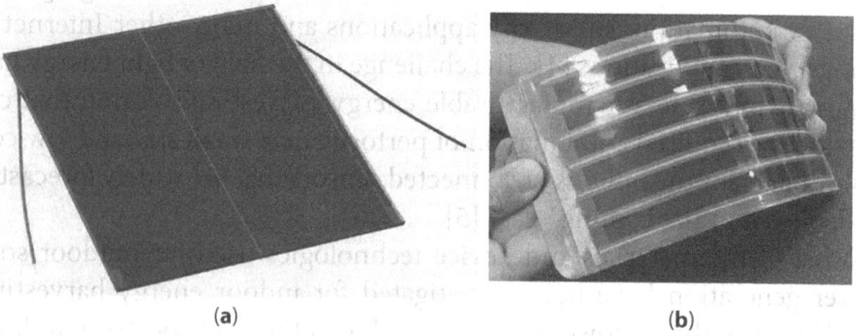

Figure 8.1 A silicon PV energy harvester (Panasonic) [10] (a). Dye-sensitized solar cell (DSSC) PV energy harvester (G24 Power Ltd.) [13] (b).

Dye-sensitized solar cells (DSSC), perovskites [7] and organic photovoltaic (OPV) [8] offer a power density increase (~7 μW/cm^2 at 200 lux) and can be made flexible (Figure 8.1 (b)).

Dye-sensitized solar cells (DSSC) are composed of a porous layer of titanium dioxide nanoparticles, covered with a molecular dye that absorbs sunlight. DSSCs work in low-light conditions and offer potential for indoor use. The major disadvantages of DSSC are the use of a liquid electrolyte (which can have temperature stability problems), and costly ruthenium (dye) and platinum (catalyst) materials used. Despite over 10 years of intense R&D on DSSCs their long-term stability and power output fall short of the requirements for many practical applications.

Perovskite solar cells have achieved good efficiency levels (>20% outdoors, up to 15% indoors) and whilst stability is improving, the required investment into scale-up and mass-production technologies make deployment in products unlikely at least in the short term.

Organic Photovoltaics (OPV), like DSSCs, is in principle suitable for the lower intensities experienced inside buildings. However, despite intense R&D efforts by industry and academia, long-term stability is reliant on effective encapsulation to prevent degradation of the OPV active material. The power output per area is comparable to that of DSSCs.

A general description of advantages and disadvantages of each technology is presented in Table 8.1.

8.2.2 Commercial PV Energy Harvesting Devices

Some of the PV energy harvesting devices which are or were commercially available, and their operating characteristics are shown in Table 8.2.

8.2.3 Recent Research Results for PV Energy Harvesting Devices

Aside from commercially available photovoltaic energy harvesters, academic research has been proficient in the field of light energy harvesting, often achieving higher performance than commercial PV devices. Much of the research activity is currently focusing on perovskite materials that show good promise for use in indoor environments, and to a lesser extent on DSSCs and organic solar cells. Table 8.3 below summarizes some of the most noticeable research results. It should be noted that it can be difficult to compare results from different laboratories. Due to the current lack of measurement standards for indoor solar cells, they are often characterized under different test conditions (e.g., different illumination levels, spectra, temperature and for different active area sizes).

Table 8.1 Energy harvesting cell technologies.

Cell technology	Advantages	Disadvantages
a-Si	• Cheap (<0.1 $/cm^2$) • Longer lifetime (up to ~10 years) than DSCCs and OPV • Good overall stability (aside from Staebler–Wronski effect) • Sufficient performance (~5% efficiency at >200 lux) for many simple sensing applications • Can be made flexible (but not generally available as such)	• Very poor performance at <100 lux and not particularly good at <200 lux • Low efficiency, leading to a greater surface area of cells • Unable to supply sufficient power for more demanding applications even in good lighting conditions (200 lux) • Degradation of efficiency with long-term exposure to direct sunlight (Staebler–Wronski effect) • Difficult to make very small (<1 cm^2) with existing manufacturing processes
DSSC and Perovskites	• Efficiencies of 5–15% indoor • Tolerant of polarized light • Better performance at low light levels than a-Si, i.e., more linear drop in power with reduced lux levels • Good performance with low angle of incidence • Can be made flexible	• Limited lifetime (around 5–10 years) • Higher cost than silicon in the short to medium term • Suitable low-cost barrier layers are still a challenge • Efficiencies not demonstrated in production so far • Voltage appears to drop off very rapidly below 50 lux • Large surface area typically required (>20 cm^2) and lower Open-circuit Voltage V_{oc} (0.58 V at 200 lux)

(Continued)

Table 8.1 Energy harvesting cell technologies. (*Continued*)

Cell technology	Advantages	Disadvantages
OPV	• Potential to be low cost with volume – easily printed in wide areas • Flexible • Low cost substrates, e.g., polyester • Translucent • Color tuning • Thin, lightweight • Ease of integration into devices • Efficiency advantages over a-Si at very low light levels (up to 50% extra at <200 lux) • Low temperature process • No exotic, rare or toxic materials required • Possibilities for tandem architecture to harvest IR or short wavelength	• Lifetime around 2–3 years with flexible encapsulations and 5–10 years in glass • Suitable low-cost barrier layers are still a challenge • Higher cost than silicon in the short to medium term • Efficiencies in labs 5–10% and not expected to exceed this for several years • Roll to roll produced cells – efficiency ~4% • Low power of approximately 3–4 μW/cm² at 200 lux and negligible below 50 lux • V_{oc} drops significantly below 200 lux • At best pilot lines but can produce modules

Table 8.2 Characteristics of commercial energy harvesting devices at 200 lux illumination, unless otherwise stated.

Manufacturer	**Total area**	**Open circuit voltage**	**Operating voltage**	**Operating current**	**Operating power**	**Technology used**	**Light operation range**
EnOcean ECS 300 [9]	35.0×12.8×1.1 mm	4 V	3 V	4.5 μA	13.5 μW	Amorphous silicon solar cell	200 lux
EnOcean ECS 310 [9]	50.0×20.0×1.1 mm	4 V	3 V	11 μA	33 μW	Amorphous silicon solar cell	200 lux
Panasonic AM-1606C-MEL [10]	15×15×0.9 mm	3.6 V	2.6 V	3.1 μA	8.06 μW	Amorphous silicon solar cell	50-1000 lux
Panasonic AM-1456CA [11]	25×10×1.3 mm	2.4 V	1.5 V	5.3 μA	7.95 μW	Amorphous silicon solar cell	50-1000 lux
Eight19 OPV (200 lux) [12]	65×110×0.8 mm	2.6 V	1.6 V	32 μA	5.12 μW	Organic photovoltaic	200-1000 lux
Eight19 OPV (1000 lux) [12]	65×110×.8 mm	3.3 V	2.3 V	180 μA	414 μW	Organic photovoltaic	200-1000 lux
GCell DSSC module [13]	200×186×1.2mm	5.11 V	3.84 V	240 μA	921.6 μW	DSSC	1000 lux

Table 8.3 Research results for state-of-the art indoor PV energy harvesters.

University/research organization	Technology	Performance	Comments
University of Swansea (UK), 2018 [14]	Perovskite	16.3–19.9 μW/cm² at 200 lux	Measured under fluorescent illumination (light spectrum not indicated)
RCNPV (Taiwan), 2019 [15]	Perovskite	11.7%–23.7% indoors	Performance depending on solar cell area
Linköping (Sweden)/ Beijing University (China), 2019 [16]	Organic	26.1% at 1000 lux	Performance depending on illumination level (worse at low light level)
EPFL (Switzerland) [17]	DSSC	27.5% at 200 lux	Performance measured based on photoactive area (not total area). Efficiency increases with light level

8.3 Lightricity's PV Energy Harvesting Technology

8.3.1 Introduction

As we have seen in the previous sections, realizing the full potential of light energy harvesting in energy-harvesting powered systems requires a new approach where device efficiency and manufacturing technology are pushed beyond the state-of-the-art limits. Lightricity has developed a high-efficiency PV energy harvesting module for efficient conversion of indoor light. These are typically artificial sources such as incandescent and fluorescent in legacy deployments, and LED in current indoor lighting applications [18].

Lightricity uses a high purity and long lifetime (>20 years) III-V material that delivers over 22 $\mu W/cm^2$ under 200 lux conditions (white LED),

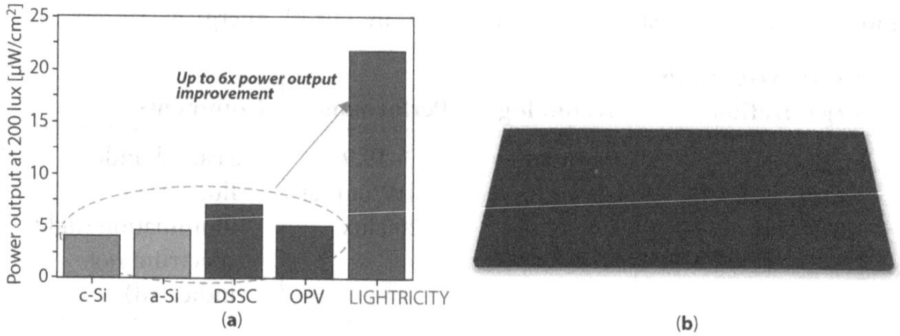

Figure 8.2 (a) Comparison of commercial indoor Photovoltaic (PV) Energy Harvesting technologies (Table 8.2 and Lightricity data, measured under 200 lux white LED illumination). (b) Lightricity PV Energy Harvesting prototype module (2 cm x 5 cm).

corresponding to 30–35% under artificial lighting (indoor fluorescent, white LED, high-pressure sodium lamps found in parking, outdoor street lighting applications) (Figure 8.2). The conversion efficiency was determined by measuring the spectra of the light sources used, as well as device quantum efficiency (measured with a Newport QE/IPC setup for solar cell characterisation). The device can also be made fully black for seamless integration into a product or environment. After scale-up, Lightricity's technology is expected to achieve cost parity (per µW) with silicon for small size devices (mm^2-cm^2).

8.3.2 Energy Harvester Device Fabrication and Device Characteristics

A prototype was successfully scaled up from a first 1 mm^2 experimental cell to a packaged 10 cm^2 multi-cell module (Figure 8.3). Independently verified module output power densities of 20.5 $\mu W/cm^2$ at 200 lux under

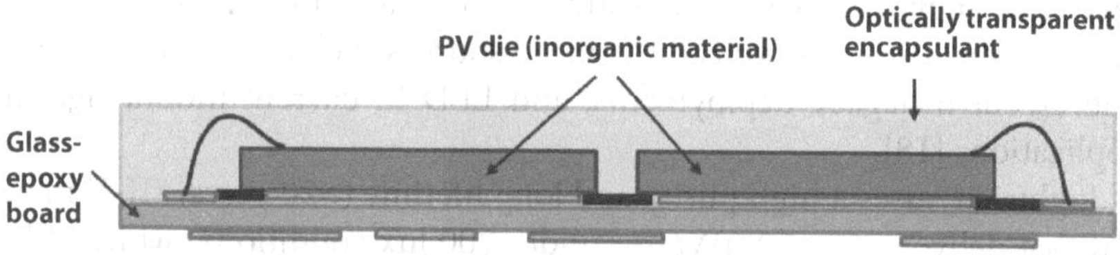

Figure 8.3 Side view of semiconductor packaging method used for Lightricity's energy harvester.

warm white LED lighting were achieved (3–6 times higher than the best commercially available products).

In addition, the technology has a number of other technical advantages, including:

- High dynamic range: The module operates with both high efficiency (>30%) and high single-cell operating voltage (>1 V) even at ultra-low light levels (<20 lux) where other technology exhibit steep roll-off in performance. The 1 V operating voltage complements emerging low voltage, power efficient chipsets for wireless communications, power management, and microcontrollers.
- High manufacturability and scalability: The technology is compatible with existing manufacturing facilities.
- Highly stable material and device performance: The module exhibits no performance degradation under indoor conditions. Devices stored under ambient temperature conditions have shown no measurable performance degradation over a period of 5 years.
- Attractive cosmetic appearance. Its black finish enhances the aesthetic impact on final product design.
- Superior outdoor performance: The module is also more efficient than low-cost crystalline silicon (16.5% efficiency under AM1.5G solar spectrum conditions). This means system integrators are able to restrict their choice of PV technology to a single option for several products covering various indoor and outdoor applications.

The open-circuit voltage range of the PV energy harvesting module with mechanical specifications as show in Table 8.4 is 4.0 ~ 4.9 V when light intensity changes from 10 ~ 1000 lux. At the indoor illumination conditions of 200 lux (white LED illumination, CCT 2,831 K), operating power

Table 8.4 Mechanical specifications.

Geometrical parameters	
Shape	Rectangular
Outer module dimensions	5.0 cm × 2.0 cm
Thickness	>1.5 mm

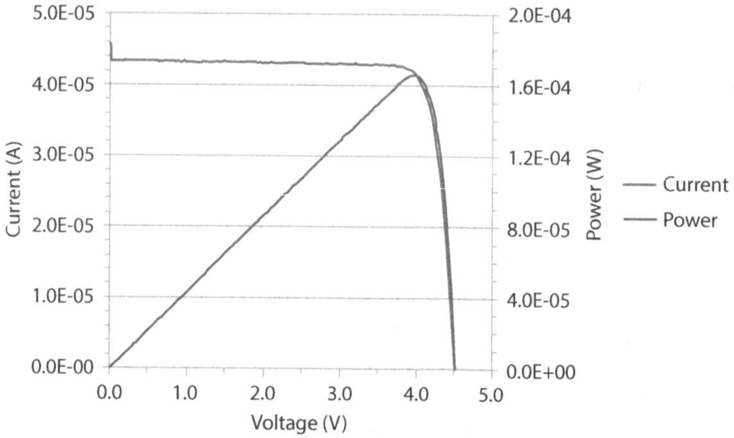

Figure 8.4 I-V (blue line) and the P-V curves (red line) of the PV energy harvesting module at 200 lux indoor illumination conditions.

at maximum power point is 170.0 μW with V_{oc} = 4.61 V and I_{sc} = 43.8 μA, corresponding to an active area power density of 21.8 μW/cm^2 (Figure 8.4).

A technical comparison of Lightricity's modules with the primary competing technology is shown in Table 8.5, and its performance as a function of illumination level and temperature in Figure 8.5.

Table 8.5 Comparison of Lightricity's modules with the primary competing technologies.

Specification at 200 lux illumination	**V_{OC} (V)**	**J_{SC} (μA/cm^2)**	**FF (%)**	**Max. output power (μW/cm^2)**
Lightricity energy harvesting cell in module	1.15	22.4	84.8	21.8
a-Si module (Panasonic Amorton) "Best in class"	0.63	17.0	65.4	7.0
a-Si module (EnOcean ECS 310)	0.5	12.5	58.9	3.7
SunPartner Wysips Crystal				~2.1 est.

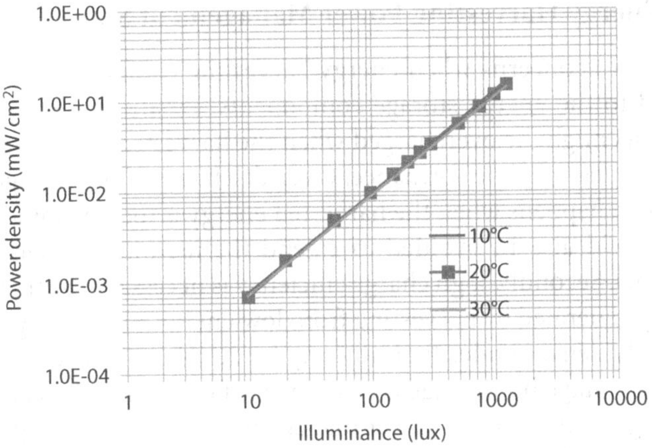

Figure 8.5 Output power density of Lightricity's energy harvesting module as a function of illumination level and temperature.

Therefore, it is apparent that PV energy harvesting can provide significant amounts of energy for IoT-based application scenarios at illumination conditions down to 200 lux and below [19].

8.4 High-Efficiency PV Energy Harvesting Power Supplies

8.4.1 Introduction

A light energy harvesting component, such as the one developed by Lightricity, is a necessary element to provide power to a IoT sensor but it is not sufficient on its own to replace directly and effectively a primary battery. The main reason is that output from the PV energy harvesting device is not regulated and will not sustain a load drawing more current than the PV energy harvesting device continuously produces (~40 μA). In IoT sensors, this situation occurs during measurement (peak currents of ~10–100 mA typically required for ~40–100 ms) and transmission (10–20 mA typically required for hundreds of ms) intervals [20]. A power management interface is thus typically required between the energy harvesting component and the load [21]. This power management interface should draw as little power as possible and should be integrated with the energy harvesting component to maximize overall system efficiency and to reduce the overall system footprint. In this section, we discuss the approach we have taken to provide efficient power management functionality.

8.4.2 Energy Harvesting Power Management Solutions

There are two concerns when energy harvesting components are directly connected to the energy storage components. Firstly, the energy storage components might be overcharged when the harvested energy is too high, resulting in degraded performance. Over-discharge of the storage components, which happens when the harvested energy is too low for a sustained period, should also be avoided to extend the lifetime of energy storage components. Secondly, high charge efficiency is necessary to utilize most of the energy generated by harvesting from low energy sources (e.g., dimmed light). Lastly, sensor nodes typically require 3.3 V or 1.8 V voltage supply, therefore a voltage boost stage is required as if only one solar cell is used.

The diagram of a typical power management circuit for energy harvesting is shown in Figure 8.6, where PV voltage is boosted to a pre-set voltage (such as 4.0 V) to charge an energy storage element, e.g., a battery. To output voltages such as 3.3 V or 1.8 V to a load circuit, a secondary buck convertor is often employed. To operate the PV device at maximum output power, an impedance matching block is added between the PV and the boost converter for tracking the maximum output power point of the PV, which changes with the illumination conditions (MPPT function).

In summary, the functions of the power management chip are to provide impedance matching, boost/buck conversion and regulated output to charge a battery/supercapacitor (also ensuring over-charge and over-discharge protection) while providing power to a load such as a sensor node.

An interesting aspect of Lightricity's energy harvesting device described earlier is that with solar cells in series, the input boost convertor may not be necessary since the output voltage of the PV module is sufficiently high (~4 V at 200 lux), which means that a simple charger optimized for low power applications can provide the functionality required.

A battery charger, LTC4071 from Linear Technology, has therefore been employed for power management. The LTC4071 IC is a Li-battery charger optimized for low power energy harvesting with over-charge (4.0~4.2 V selectable) and over-discharge (2.7~3.2 V selectable) protection, a maximum

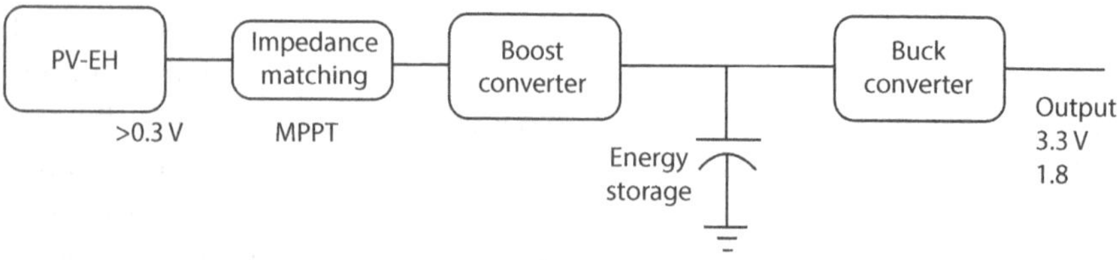

Figure 8.6 Typical circuit diagram of commercially available power management ICs.

pulse current of 400 mA at <10 ms and a continuous current of up to 50 mA. The block diagram of the chip is shown in Figure 8.7. The energy harvester is connected to R_{in}, a Li-battery or supercapacitor is connected to Bat, and the output (load) is connected to V_{cc}. When the supercapacitor voltage is lower than the over-discharge voltage, the supercapacitor is charged through R_{in} and the diode. When the supercapacitor voltage exceeds the over-discharge protection voltage, MP_1 is switched on (4–6 Ω on-resistance). The charge current will be shunted through the MP_2 path when the supercapacitor voltage is higher than over-charge protection voltage.

The output current of the energy harvesting device changes with the illumination condition, as shown in Figure 8.8. For a given illumination, the output current is almost constant when output voltage changes from 0–85% of the open-circuit voltage, which means that the output power of the energy harvester mainly changes with operating voltage. The output power of the EH changes as shown in Figure 8.9, where the maximum output power point is labeled by a circle located at ~85% of the open-circuit voltage of the device.

Lightricity's energy harvesting device provides an open-circuit voltage of 3.9 V ~ 4.8 V, which can be directly used to charge a 3.7 V battery, or a supercapacitor. That is, instead of setting operating voltage of the energy harvesting device by MPPT, the operating voltage of the energy harvesting device is directly set as the storage element voltage when the energy harvester is directly connected to the storage element.

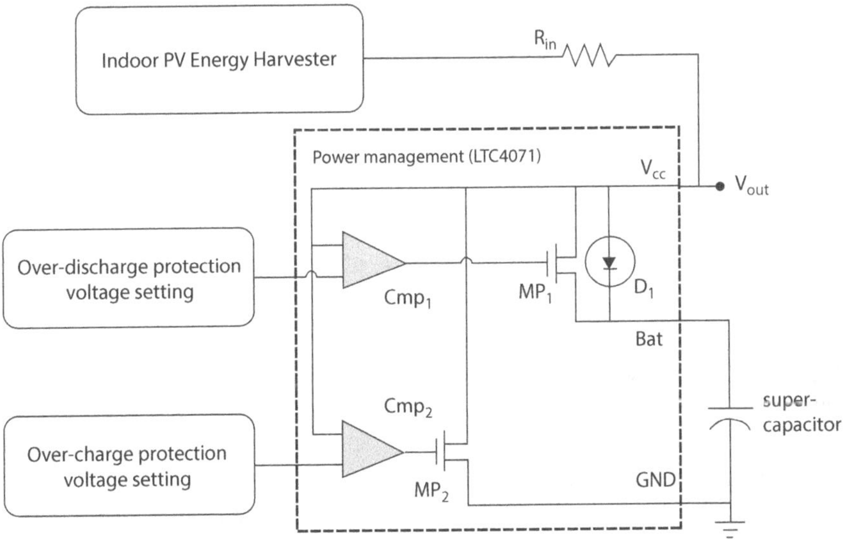

Figure 8.7 Block diagram of the LTC4071.

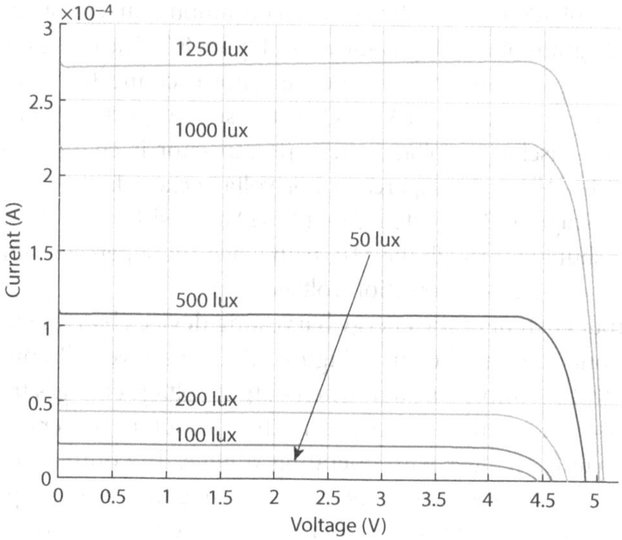

Figure 8.8 Light IV curves of Lightricity's energy harvesting device as a function of illumination level (white LED lighting).

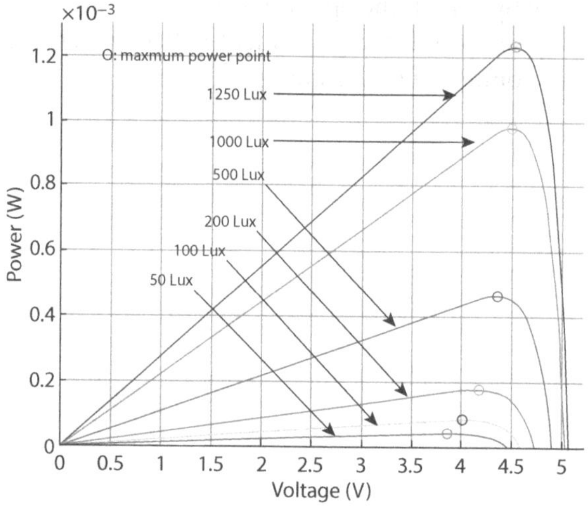

Figure 8.9 Output power of Lightricity's energy harvesting device as a function of illumination level (white LED lighting).

A charging chip such as the LTC4071 has been found to perform well without MPPT. This can be explained from the characteristic curves of the energy harvesting device tested under different illumination conditions. Figure 8.9 shows the output power of the solar panel with its maximum output power point. Figure 8.10 shows the normalized efficiency when maximum output power is treated as 100%, demonstrating >80% efficiency if the operating voltage of the solar panel is set in the region of 3.6 V to 4.2 V. In the region of 3.5 V to 4.0 V, the lower the illumination, the higher the efficiency. This property is advantageous under very low light level conditions.

The reason that this power management approach achieves high charge efficiency is counter-intuitively related to the fact that it does not use MPPT. Total efficiencies of 80% are not usually achieved in commercially available power management chips with MPPT [23]. Referring to Figure 8.6, assuming the efficiencies of the boost and buck converters are 80%, the total efficiency of the power management chip becomes as low as 64%. Furthermore, MPPT circuits contribute to quiescent current which decreases the charge efficiency further, especially under low illumination condition where harvested energy might be even lower than the energy required for running these circuits.

We have tested a power module which is composed of Lightricity's energy harvesting device, a LTC4071, a Li-battery or supercapacitor with a switched resistive load to simulate the power consumption of a wireless sensor node during the measurement and data transmission periods, and during the sleep period. This simulated environment integrates practical considerations of the quiescent current of the LTC4071, the charge efficiency and the leakage current, to evaluate if the energy harvested is sufficient to, for example, autonomously power a building automation sensor node.

Figure 8.11 shows the experimental setup, where the LTC 4071 is connected to the PV energy harvesting module and a simulated load. The low leakage switch used is a Vishay DG467, which has maximum on-resistance

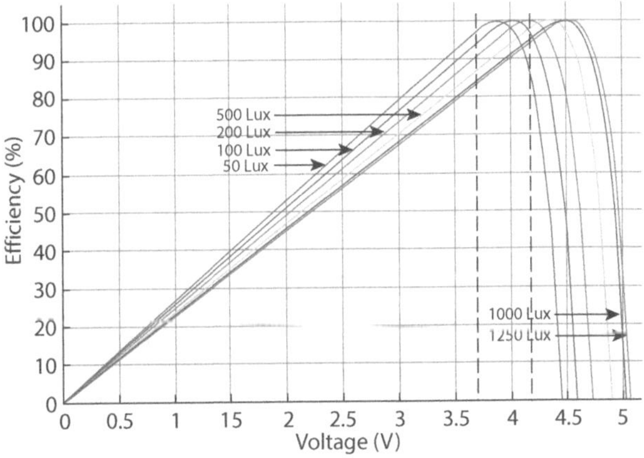

Figure 8.10 Normalized efficiency of Lightricity's energy harvesting device as a function of operating voltage.

of 10 Ω, fast switch time of 100 ns and a maximum leakage current of 10 nA. The switch is controlled by a pattern signal generator which produces a 15 ms pulse signal every 200 seconds. The load resistor is set to 100 Ω. With a pre-charged battery of 3.7 V, the total discharge of the battery within a measurement period has been measured to be 3.7 V/110 Ω * 15 ms = 504 μAs, which is higher than the 400 μAs required by the building automation CO_2 sensor discussed in the next section.

8.4.3 System Integration and Performance Testing

This section presents the development of an Indoor PV Energy Harvesting Power Module (IPEHPM) for wireless sensor nodes, using the approach described in the previous section. The sensor node utilizes a low-power consumption autonomous CO_2, temperature and humidity sensor and its associated signal conditioning circuits from GSS (Gas Sensing Solutions Ltd [22]), powered using the newly developed IPEHPM. The energy harvested using the IPEHPM provides a high open-circuit voltage, directly stored in a supercapacitor without using a voltage regulator and accordingly without using maximum power point tracking (MPPT). Storage efficiency analysis and efficiency testing of the IPEHPM are described to demonstrate that high energy storage efficiency can be achieved for typical indoor illumination conditions of 200 lux. Testing of the supercapacitor in-circuit self-discharge current is reported since the data sheet self-discharge current of the supercapacitor is typically defined in the steady state only. Dry testing of the IPEHPM at a fixed illumination of 200 lux to validate the functions of IPEHPM, such as overcharge protection, over-discharge protection and reactivation after over-discharge

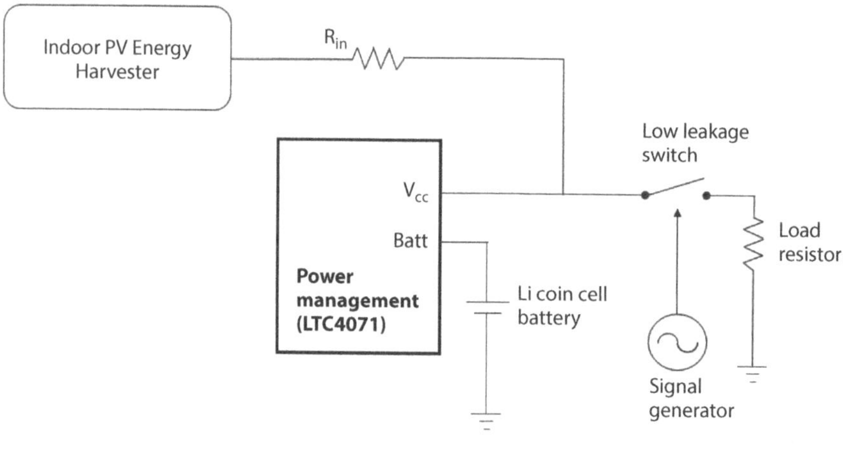

Figure 8.11 Experimental setup for load simulation measurements.

protection, was also carried out. Efficiency measurements of the IPEHPM are reported, too. Powering the IoT gas sensor node with the IPEHPM to measure CO_2 concentration at 200 lux indoor lighting for automatic building air quality control is reported in Section 8.5.2: Wireless Building Automation Sensors.

The IPEHPM has been designed and fabricated as shown in Figure 8.12. A low voltage drop Schottky diode has been added between the PV energy harvester output and R_{in} to protect the PV energy harvester being charged by the storage when illumination conditions are poor. The over-charge protection voltage can be set as 4.0, 4.1 or 4.2 V on the board. The over-discharge protection voltage has been set as 3.2 V. The connector located on the right part of the PCB is for power output and also provides an additional fast charging access to the supercapacitor. All electronic components are mounted on the bottom layer of the PCB. The PV energy harvesting module, which has the same dimension as the PCB, has been soldered on the top layer of the PCB. The fabricated IPEHPM has a dimension of 50 mm × 20 mm × 15 mm.

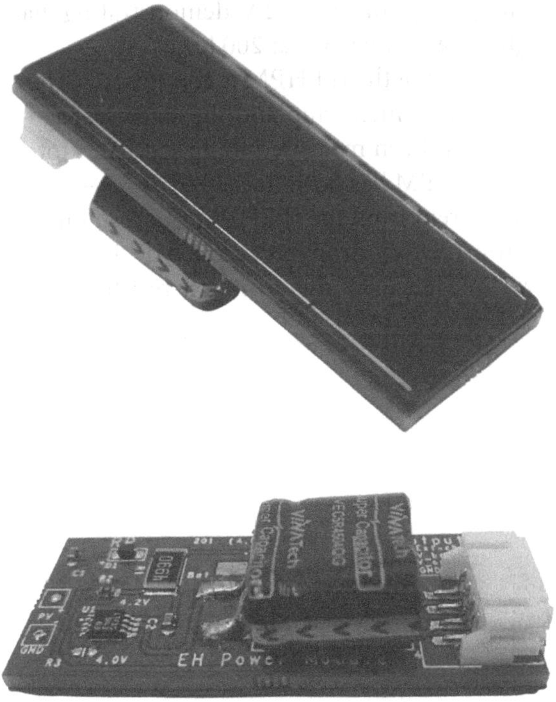

Figure 8.12 The 50 mm × 20 mm × 15 mm IPEHPM. Top: PV energy harvesting module side of the IPEHPM. Bottom: electronic components side of IPEHPM.

Storage efficiency of the IPEHPM has been examined using the designed PCB by replacing the energy storage supercapacitor with a Keithley source-meter acting as a voltage source, so that the charge current flowing into the storage can be measured at different operating voltages when the PV energy harvester is illuminated at 200 lux. Varying the voltage of the Keithley source-meter from 3.6 V to 4.2 V with a 0.1 V voltage step, the current flowing into the Keithley, together with the PV output current and the output voltage, has been recorded at the same time.

Figure 8.13 shows that the maximum output power of the PV energy harvesting module is 180.95 μW when the output voltage of the PV energy harvester is at 4.0 V, while the maximum power the storage received is at the voltage of 3.8 V. The 0.2 V difference is caused by the voltage drop measured across the protection Schottky diode.

Power output efficiency of the power module has been calculated by dividing the storage received power by the maximum PV energy harvesting module output power, as shown in Figure 8.13. The achieved maximum storage efficiency is 93.1% at the storage voltage of 3.8 V while the minimum is 79.6% at 4.2 V. The average storage efficiency is 88.7% for the tested storage voltage range of 3.6 ~ 4.2 V, demonstrating that the IPEHPM has achieved high storage efficiency at 200 lux.

The quiescent current of the IPEHPM is composed of the quiescent current of the power management chip and the static leakage current of the supercapacitor, and has been measured to be 1.2 μA in total. The resistive impedance of the IPEHPM has been measured as 5.5 +/−1.0 Ω. Therefore, outputting a high current from the IPEHPM can result in a relatively high output voltage drop (mainly dropped across MP_1 in the power management chip shown in Figure 8.7 due to its on-resistance). For low-power

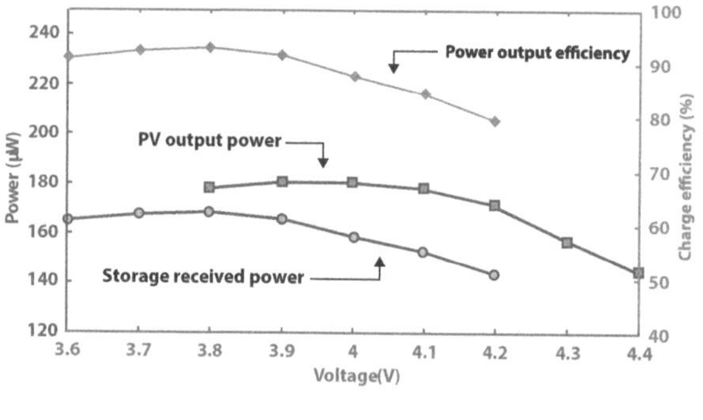

Figure 8.13 Output power of the PV energy harvesting module (pink squares), storage received power (yellow circles) and the storage efficiency (green diamonds).

applications such as the ~20 mA current pulses encountered for low-power wireless transmission, the maximum output voltage drop in active mode is less than 130 mV, which is insignificant (4% of the supply voltage of 3.3 V).

8.5 Applications of Light Indoor Energy Harvesting

As we have seen in the previous sections, energy harvesting can be combined with low-power electronic circuitry and storage to provide a complete energy harvesting solution for all use cases requiring sensing, information-storage or transmission. We will discuss some of the use cases with significant mass-market potential.

8.5.1 Watches and Wearable Devices

One of the promising applications of high-efficiency light energy harvesting technology is in wearable devices. Commercially available light energy harvesting devices to date have been almost exclusively based on amorphous silicon technology, with a-Si providing output powers of ~5 μW/cm^2 under typical indoor lighting conditions (200 lux). This is sufficient to autonomously power the quartz movement in a traditional solar watch (~5 μW average power consumption), or a solar watch or solar calculator with segmented LCD display (~10–15 μW average power consumption). Indeed, solar powered watches and calculators have been the earliest mass-market applications of indoor light energy harvesting technology.

Using high-efficiency light energy harvesting devices would increase significantly the amount of electrical power available to watches and wearable devices, offering the possibility of adding "smart functionality" (e.g., BLE connectivity, health sensors) to the watch, and the possibility of adding a more performant display such as a low-power memory-in-pixel (MIP) reflective display (Figure 8.14).

8.5.2 Wireless Building Automation Sensors

Energy saving smart air quality control in buildings can be achieved via automated control of air handling systems based on the real-time measurement of air quality parameters, such as CO_2 concentration, temperature and humidity, using ambient energy harvesting powered IoT nodes. The environmental parameters change slowly so the air quality parameters inside a building need to be measured periodically rather than continuously. The power that the ambient energy harvester can produce might be lower than that required for

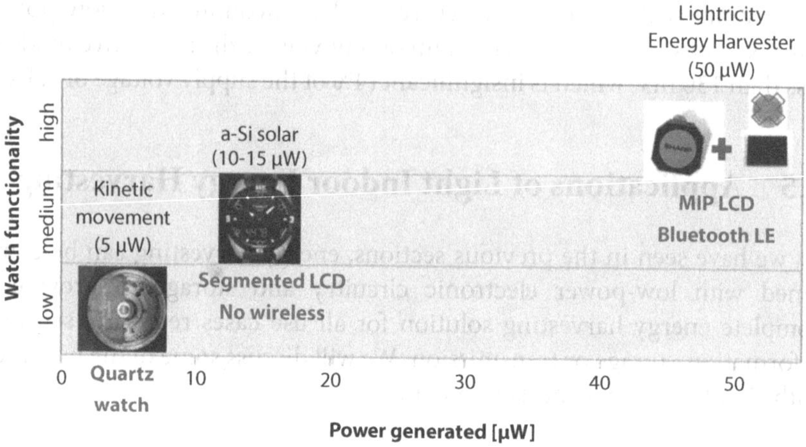

Figure 8.14 Functionality vs. power requirements in energy-harvesting powered watches.

a measurement, but the harvested energy can be continuously accumulated into energy storage components so that a high-power pulse can be supplied for short-term measurement and wireless transmission events.

For building automation applications, the harvested PV energy can be stored in a rechargeable Lithium battery or a supercapacitor. Both energy storage devices are suitable for powering low-power building automation nodes in terms of the storage capacity. Millions of charge cycles may make supercapacitors a better solution for lifetime "fit and forget" applications; on the other hand, their higher self-discharge is a concern for their application in low-power IoT devices [24].

We utilize the energy harvesting power supply module (IPEHPM) described in Section 8.4: High-Efficiency PV Energy Harvesting Power Supplies, to power the sensor node, with a 0.5 F supercapacitor from VinaTech. The storage supercapacitor provides 3.6 V to 4.0, 4.1 or 4.2 V with a peak pulse current of up to 100 mA, for up to 600 ms, to the building automation sensor. The building automation sensor is comprised of a low-power CO_2 sensor, a microcontroller and a wireless radio, as shown in Figure 8.15. During a measurement, the sensor is powered on to measure CO_2 and to then send the measured value wirelessly to a gateway (active mode, labeled as T_1 in Figure 8.16). Subsequently, sensor and radio are powered off to enter sleep mode (labeled as T_2 in Figure 8.16), during which time harvested energy is accumulated for the next measurement.

The current profile of the GSS gas sensor for air quality measurements is shown in Figure 8.17, where the first peak is for sensor power on, followed by the second peak for measurement and the last peak for wireless data

Figure 8.15 A 70 × 50 × 20 mm autonomous IoT sensor node for building automation. On the left is the PV energy harvesting power supply, on the right the CO_2 sensor, microcontroller and radio module.

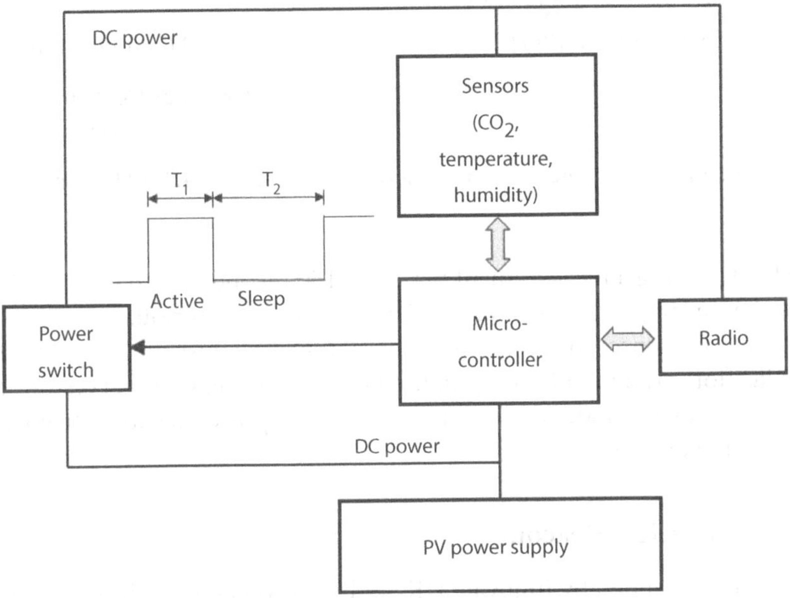

Figure 8.16 Block diagram of the building automation sensor. Sensor and radio are only powered during active mode (T_1).

transfer. All current pulses are within the IPEHPM specifications of 100 mA at 600 ms. The required current to maintain the microcontroller in sleep mode is 4 μA. The total discharge for a measurement and transmission event every 2.5 mins is calculated as ~4.3 mC.

As shown in Table 8.6, the effective charge current of the IPEHPM is 38.6 μA at 200 lux, i.e., the total charge the PV energy harvester provided is

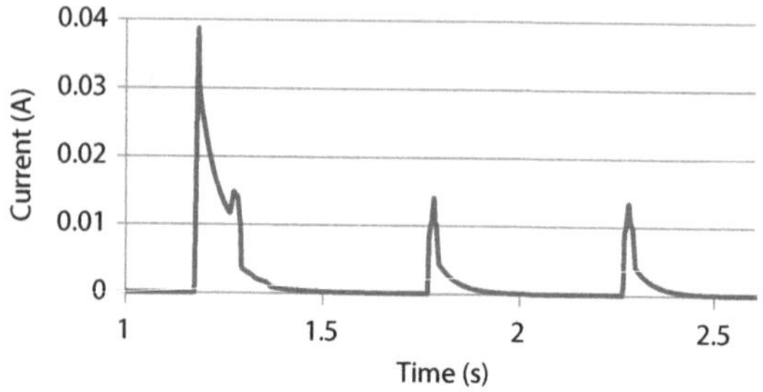

Figure 8.17 Current profile for a single CO_2 measurement and wireless transmission event.

Table 8.6 Summary of the power generated and provided by the IPEHPM for CO_2 concentration measurements every 2.5 mins at 200 lux (power in μW).

PV energy harvester output power	159.7
Available power from IPEHPM	38.6 × 3.6 = 139.0 (min)
	38.6 × 4.2=162.1 (max) 150.5 (average)
Power consumed by the CO_2 sensor node	28.3 × 4.2 = 118.9 (max)

5640 μC during a measurement period of 150 second, which is larger than the charge consumption of 4240 μC during the same period.

The CO_2 sensor node for building automation provides a successful application example of using high-efficiency light energy harvesting for powering an IoT node which could not be easily powered with light energy harvesting otherwise.

8.5.3 Wireless Beacons

In a project funded by Innovate UK, and in partnership with Ilika (developing solid-state battery storage) a second generation of compact everlasting power supply that can connect to any IoT sensor or wireless tracking devices and beacons has been developed (Figure 8.18). The power pack has been demonstrated to autonomously power a Bluetooth low-energy (BLE) beacon, with applications in asset tracking in warehouses and hospitals, logistics and location-dependent services.

In a second, final design the beacon was miniaturised to minimal size (35 × 25 mm). With its thin form factor (<4 mm), the final device resembles a thin patch. The final device comprises a Lightricity 25 × 10 mm light

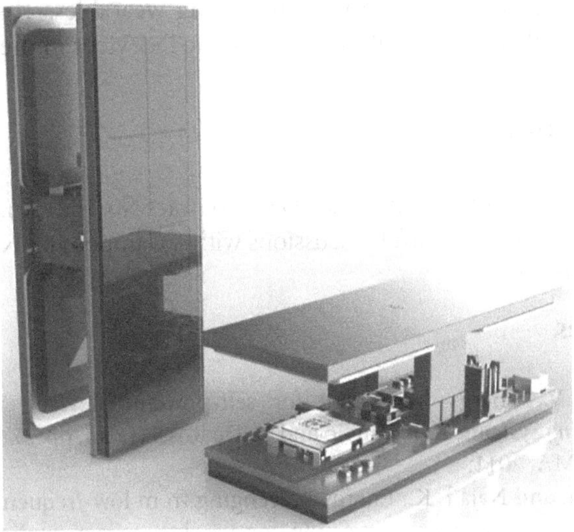

Figure 8.18 Energy harvesting power pack (Lightricity, Ilika).

energy harvesting device, an Ilika 250 μAh Stereax M250 solid state battery, a Nordic nRF52832 Bluetooth Low Energy SoC with built-in temperature sensor and power management. The perpetual beacon measures temperature at regular intervals and transmits the data using Bluetooth Low Energy to a smartphone app. The app displays temperature information as well as the battery's state of charge. This device replicates non-critical sensors where the data could be transmitted wirelessly to a hub instead of smartphone. The minimal size and maintenance-free operation of the beacon enables installation in difficult-to-reach places in many use cases, e.g. industrial IoT, asset tracking and automotive.

8.6 Summary and Concluding Remarks

Powered using energy harvesting: this has been widely considered the only logical and sustainable solution for providing power to IoT devices in the medium to long term. The developments summarized in this chapter demonstrate that powering IoT devices and sensors by energy harvesting is both practical and beneficial for lowering cost of ownership. New energy harvesting and sensor technologies are now making it a viable, practical option for real-world deployments in the short term, too. It is also in line with current trends of device miniaturization and low-power consumption. As a consequence, there are now commercially available energy harvesting solutions which can provide enough energy and peak power for a wide range of wireless applications, including wearables. This opens the way to large-scale deployment of connected IoT devices in the short term.

Acknowledgments

This work was supported by Innovate UK (Contract No. 102156). The authors gratefully acknowledge fruitful discussions with J. Campos and N. Evans.

References

1. Leonov, V., Energy harvesting for self-powered devices, in: *Wearable Monitoring Systems*, A. Bonfiglio and D. De Rossi (Eds.), pp. 27–49, Springer, Boston, MA, 2011.
2. Kulah, H. and Najafi, K., Energy scavenging from low-frequency vibrations by using frequency up-conversion for wireless sensor applications. *IEEE Sensors J.*, 8, 261–268, 2008.
3. Drayson Technologies, RF Energy Harvesting for the Low Energy Internet of Things, Drayson Technologies Ltd, London, UK, 2015.
4. DeFeo, C., Energy harvesting and the internet of things, in: *Green Information Technology*, M. Dastbaz, C. Pattinson, B. Akhgar (Eds.), pp. 151–160, Elsevier, Waltham, MA, 2015.
5. Al-Fuqaha, A., Guizani, M., Mohammadi, M., Aledhari, M., Ayyashet, M., Internet of Things: A survey on enabling technologies, protocols, and applications. *IEEE Commun. Surveys Tutorials*, 17, 2347–2376, 2015.
6. Foti, M., Tringali, C., Battaglia, A., Sparta, N., Lombardo, S., Gerardi, C., Efficient flexible thin film silicon module on plastics for indoor energy harvesting. *Sol. Energy Mater. Sol. Cells*, 130, 490–494, 2014.
7. Lucarelli, G., Di Giacomo, F., Zardetto, V., Creatore, M., Brown, T.M., Efficient light harvesting from flexible perovskite solar cells under indoor white light-emitting diode illumination. *Nano Res.*, 10, 2130–2145, 2017.
8. Mori, S., Gotanda, T., Nakano, Y., Saito, M., Todori, K., Hosoya, M., Investigation of the organic solar cell characteristics for indoor LED light applications. *Jpn. J. Appl. Phys.*, 54, 071602, 2015.
9. EnOcean, EnOcean Self-powered IoT - ECS 300, EnOcean GmbH, Munich, Germany, [Online], https://www.enocean.com/en/enocean-modules/details/ecs-300/.
10. Panasonic, Panasonic-AM-1606C-Datasheet, Panasonic Solar Amorton Co., Ltd, Tokyo, Japan, [Online], https://www.mouser.com/ds/2/315/panasonic_AM-1606C-1197092.pdf.
11. Panasonic, Panasonic-1456CA Datasheet, Panasonic Solar Amorton Co., Ltd, Tokyo, Japan, [Online], https://mouser.com/datasheet/2/315/panasonic_AM-1456CA-1196920.pdf.

12. Eight19, Indoor engineering sample specification sheet previously available. Eight19 Ltd, Cambridge, UK, [Online], https://eight19.com/wp-content/uploads/2016/08/Indoor-Engineering-Sample-Specs-Sheet-E19_ES_7_12_002.pdf.
13. GCell, GCell sample DSSC module, G24 Power Ltd, Newport, UK, [Online], https://gcell.com/product/gcell-sample-dssc-module.
14. Lee, H., Barba, J., Meroni, S.M.P., Du, T., Lin, C., Pockett, A., Troughton, J., Jain, S.M., Rossi, F.D., Baker, J., Carnie, M.J., McLachlan, M.A., Watson, T.M., Durrant, J.R., Tsoi, W., Outstanding indoor performance of perovskite photovoltaic cells – Effect of device architectures and interlayers. *Solar RRL*, 3, 1800207, 2019.
15. Chuang, B. and Hwang, A., *Taiwan researchers develop perovskite solar cells to convert indoor light*, DIGITIMES, Taipei, Taiwan, 2019, [Online], https://www.digitimes.com/news/a20190417PD210.html, accessed 01 November.
16. Cui, Y., Wang, Y., Bergqvist, J., Yao, H., Xu, Y., Gao, B., Yang, C., Zhang, S., Inganäs, O., Gao, F., Hou, J., Wide-gap non-fullerene acceptor enabling high-performance organic photovoltaic cells for indoor applications. *Nat. Energy*, 4, 768–775, 2019.
17. Cao, Y., Liu, Y., Zakeeruddin, S.M., Hagfeldt, A., Direct contact of selective charge extraction layers enables high-efficiency molecular photovoltaics. *Joule*, 2, 1108–1117, 2018.
18. Montoya, F.G., Peña-García, A., Juaidi, A., Manzano-Agugliaro, F., Indoor lighting techniques: An overview of evolution and new trends for energy saving. *Energy Build.*, 140, 50–60, 2017.
19. Savanth, A., Bellanger, M., Weddell, A., Myers, J., Kauer, M., Energy neutral sensor system with micro-scale photovoltaic and thermoelectric energy harvesting. *J. Phys. Conf. Ser.*, 1052, 012069, 2018.
20. Bernegger, J. and Meli, M., Comparing the energy requirements of current bluetooth smart solutions, in: *Embedded World: Exhibition & Conference, Nuremberg*, February 25–27, 2014.
21. Weddell, A.S., Merrett, G.V., Al-Hashimi, B.M., Photovoltaic sample-and-hold circuit enabling MPPT indoors for low-power systems. *IEEE Trans. Circuits Syst.*, 59, 1196–1204, 2012.
22. Gibson, D. and MacGregor, C., A novel solid state non-dispersive infrared CO2 gas sensor compatible with wireless and portable deployment. *Sensors*, 13, 7079–7103, 2013.
23. Wang, Y., Liu, Y., Wang, C., Li, Z., Sheng, X., Gyu Lee, H., Chang, N., Yang, H., Storage-less and converter-less photovoltaic energy harvesting with maximum power point tracking for internet of things. *IEEE Trans. Comput. Aided Des. Integr. Circuits Syst.*, 35, 173–186, 2016.
24. Diab, Y., Venet, P., Gualous, H., Rojat, G., Self-discharge characterization and modeling of electrochemical capacitor used for power electronics applications. *IEEE Trans. Power Electron.*, 24, 510–517, 2009.

9

Indoor Photovoltaics Based on AlGaAs

Jamie Phillips[1]*, Eunseong Moon[1] and Alan Teran[2]

[1]*Department of Electrical Engineering and Computer Science, University of Michigan, Ann Arbor, Michigan, USA*
[2]*Boston University School of Law, Boston, Massachusetts, USA*

Abstract
Photovoltaics based on III-V compound semiconductors offer high power conversion efficiency associated with a wide range of attainable bandgap energies, high optical absorption strength, excellent charge carrier transport properties, and low non-radiative recombination pathways. GaAs and AlGaAs alloys offer a mature materials technology that can offer high power conversion efficiency that is well matched to the narrow indoor lighting spectrum, with high performance under low-flux illumination conditions. Two key regimes are identified for indoor photovoltaics: 1) large-area photovoltaics for general power generation, and 2) small-area photovoltaics for self-powered miniature systems. Large-area GaAs photovoltaics based on epitaxial liftoff show great promise as a cost-competitive technique, though application to indoor power generation has not been explored in detail. Small-area (approximately 1 mm^2) photovoltaics offer an efficient energy harvesting technology that is sufficient for perpetual operation of low-power systems. AlGaAs photovoltaic cells offer among the highest reported efficiencies (>20%) for small-area devices under indoor conditions, with performance currently limited by perimeter sidewall recombination as cell area decreases. Furthermore, monolithic photovoltaic modules with series-interconnected cells offer a means to provide sufficient voltage for direct battery charging, and bypassing needs for voltage up-conversion circuitry.

Keywords: Compound semiconductors, epitaxy, non-radiative recombination, shunt resistance, internet of things, voltage up-conversion

**Corresponding author*: jphilli@umich.edu

Monika Freunek Müller (ed.) Indoor Photovoltaics: Materials, Modeling and Applications, (241–272)

9.1 Importance of AlGaAs for Indoor Photovoltaics

Indoor lighting, from either the sun or artificial lighting, can provide a key source for harvesting energy for general power generation or to independently operate standalone devices. Sunlight is more than adequate for substantial power generation, providing 1 mW/mm^2 of power under full sun, or ~100 $\mu W/mm^2$ with a conversion efficiency of 10%. However, stray sunlight is not available in all locations, or at all times. Artificial indoor lighting may also provide sufficient energy, though the intensity and spectral content are significantly different than sunlight. Today, efficient indoor artificial lighting sources, such as light-emitting diodes (LEDs) and fluorescent lamps, provide a relatively narrow band of light in the visible spectral region with a power density on the order of 1 $\mu W/mm^2$ (for an illuminance of ~600 lx). In this chapter, we will focus on photovoltaic (PV) energy harvesting from indoor artificial lighting sources, and use the term indoor lighting to specifically represent this scenario.

The optical spectrum for indoor lighting is primarily contained in a narrow spectral band in the visible region. The spectrum can also vary depending on the light source (e.g., incandescent, fluorescent, and LED) and color rendering (e.g., warm white and cool white). Fluorescent and LED lighting, in particular, have similar spectral content such that the photovoltaic response will be very similar in both cases. Further details on the spectrum of artificial indoor lighting sources are described in Chapter 5: Characterization and Power Measurement of IPV Cells. The optimal bandgap energy for a photovoltaic cell for indoor lighting will correspond to the edge of the visible spectrum, and will be larger than the optimal bandgap energy for solar illumination due to the absence of spectral content beyond the visible region. The larger bandgap energy will provide a larger voltage output while also absorbing the full spectrum provided from indoor lighting. Theoretical calculations suggest an optimal bandgap energy of 1.9–2 eV for indoor lighting sources [1] in comparison with the Shockley–Queisser limit of 1.34 eV for AM 1.5 solar illumination [2]. Silicon, with bandgap energy of 1.1 eV corresponding to the IR spectral region, is, therefore, smaller than desired for efficient energy harvesting under indoor lighting conditions. Alternatively, III–V compound semiconductor materials, such as GaAs, provide larger bandgap energy with proven high conversion efficiency and a wide range of accessible spectral windows. In addition to the bandgap energy of the material, which governs the fundamental absorption properties, output voltage, and dark current limitations, the parallel (shunt) resistance for a photovoltaic cell will have strong influence on performance under low-light conditions [1, 3–6].

We calculated the maximum energy harvesting efficiency under indoor lighting conditions based on detailed balance theory assuming photovoltaic cells with full optical absorption for energy above the material bandgap, radiatively limited performance, and full collection of photogenerated charge carriers. Three indoor lighting spectra were considered: 1) fluorescent source with daylight color rendering; 2) LED with phosphor; and 3) three-color LED, with spectra given in [7]. It should be noted that indoor lighting intensity is typically reported in photometric units of lumens or lux (lx), rather than radiometric units of watts or watts per unit area, respectively. Lighting conditions will be described in terms of lux throughout while also providing corresponding values of radiometric power density for comparison with values typically quoted for solar photovoltaics. Of the three spectra, the maximum efficiency can be obtained for a three-color LED corresponding to bandgap energy of 1.9 eV with an efficiency of 60%. Analysis to compare limiting efficiencies with variable indoor lighting sources has similarly been conducted in Ref. [1], where an optimal bandgap energy of 1.9–2 eV was determined. Higher efficiency may be obtained for indoor lighting sources in comparison with sunlight, since the narrower spectral band mitigates transparency losses and thermalization losses associated with the broadband solar spectrum. It should be noted that the conversion efficiency near the peak is relatively independent of the indoor lighting source. While higher conversion efficiency may ultimately be achieved for indoor lighting relative to sunlight, the actual light intensity is significantly lower (1–2 W/m^2).

The calculated current density versus voltage and power density versus voltage are shown in Figure 9.1. for select bandgap energies. The selected bandgap energies are those of Si, GaAs, $Al_{0.2}Ga_{0.8}As$, and $In_{0.5}Ga_{0.5}P$ as viable options for indoor photovoltaic technologies. All four materials have similar current densities, greater than 600 nA/mm^2, since they absorb all of the photons from the indoor spectra. The larger bandgap materials have larger operating voltages which result in larger maximum power densities of up to 800 nW/mm^2 at 500 lx illumination. The dependence of power density on indoor lighting illuminance is perhaps the most relevant figure for mm-scale systems. A comparison of maximum achievable power density versus illuminance for select bandgap energies is shown in Figure 9.2. For these relatively standard indoor lighting conditions, a power density > 100 nW/mm^2 can be obtained. This analysis underscores the large improvements in energy harvesting that may be achieved by considering materials with larger bandgap energy than silicon. The analysis described assumes perfect optical absorption and carrier collection, and does not consider material-specific parameters and how they would influence device design and conversion efficiency.

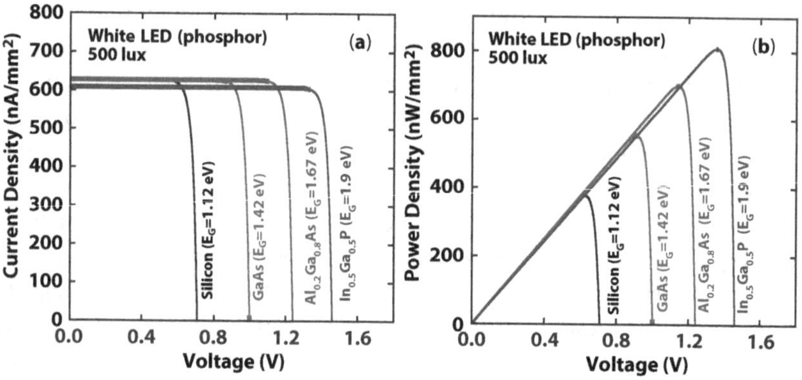

Figure 9.1 Calculated ideal (a) current density versus voltage and (b) power density versus voltage for various semiconductor materials under 500 lx illumination from a white LED.

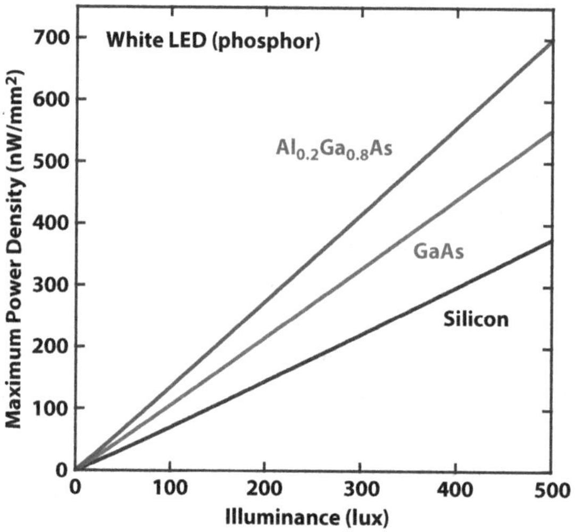

Figure 9.2 Calculated maximum power density versus illuminance for select materials under white phosphor LED illumination, illustrating that target goals of >100 nW in a 1 mm^2 area may be reasonably achieved.

In this chapter, we focus on AlGaAs photovoltaics for indoor energy harvesting. GaAs-based technology is perhaps the most advanced compound semiconductor technology based on decades of research and development and increasing market growth for optoelectronics and radio-frequency (RF) electronics. The mobile handset industry spurred major growth in GaAs technology, and is expected to play a major role for the RF front end to enable 5G [8]. Photonics are expected to further drive growth in the GaAs market, including areas in data communications and 3D sensing for smartphones. The advanced material technology, and close match of material properties for the spectral content of indoor lighting make GaAs and associated AlGaAs alloys attractive for photovoltaic energy harvesting. $Al_xGa_{1-x}As$ alloys with $x \leq 0.3$ have been previously pursued in the design of one of the top cells in multi-junction solar cells [9–11], heterostructure solar cell designs [12], and broad-spectrum intermediate band solar cells [13]. For indoor applications, AlGaAs-based photovoltaics target a narrow spectral band corresponding to artificial lighting conditions, with differing design parameters than solar cells. Applications can generally be dived into categories of either large-area cells/modules (larger than 1 cm^2) or small-area devices ($\leq$ 1 cm^2) to harvest energy for portable or wireless devices.

9.2 Design Consideration for AlGaAs III-V Photovoltaic Cells

High efficiency AlGaAs PV cells can be obtained using conventional p-n junction architectures consisting of a wide bandgap window layer, heavily doped emitter, and lightly doped base. Layers to achieve low resistivity electrical contacts and a back surface field to reduce recombination are also typically incorporated. PV cells use epitaxial techniques such as molecular beam epitaxy or metal-organic vapor phase epitaxy, where the layer design and achievement of high-quality materials is critical in achieving high conversion efficiency. An example PV cell design is described in Table 9.1, with considerations for choice of parameters described in the following.

Table 9.1 Example device structure for an AlGaAs PV cell for indoor energy conversion.

Layer	Material	Thickness (nm)	Doping (cm^{-3})
Contact	GaAs	>100	$p^{++} > 10^{19}$
Window	$Al_xGa_{1-x}As$, $x \sim 0.8$–0.9	30	$p^{+} \sim 10^{18}$
Emitter	$Al_xGa_{1-x}As$, $x < 0.45$	200	$p^{+} \sim 10^{18}$
Base	$Al_xGa_{1-x}As$, $x < 0.45$	500–3000	$n \sim 10^{17}$
Back surface field	$Al_xGa_{1-x}As$, $x > x_{Base}$	100	$n^{+} \sim 10^{18}$
Contact	GaAs	>100	$n^{+} \sim 10^{18}$

9.2.1 Base/Absorber

The primary design factor determining energy conversion efficiency for AlGaAs PV cells operating under indoor lighting conditions is the specification of the base/absorber layer of the cell. The bandgap energy of the absorber should be maximized to provide the highest voltage output, while providing optical absorption across the indoor lighting spectrum. The dependence of $Al_xGa_{1-x}As$ bandgap energy on alloy composition at temperature of 300 K is given by

$$E_G = 1.424 + 1.247x \quad (x < 0.45) \tag{9.1}$$

AlGaAs has a direct bandgap in this range of alloy composition, where higher aluminum fraction results in an indirect bandgap where the conduction band minimum transitions from the Γ to the X and L valleys. Full coverage of the visible light spectrum out to approximately 740 nm would translate to a bandgap energy of 1.67 eV, and $Al_xGa_{1-x}As$ alloy composition with $x = 0.2$. Optimizing the power conversion efficiency for typical indoor lighting conditions requires a bandgap energy of approximately 2.0 eV [1], requiring an $Al_xGa_{1-x}As$ alloy composition near the direct/indirect bandgap crossover of $x = 0.45$. In addition to optical absorption, the choice of alloy composition can impact efficiency according to the bulk material properties of the AlGaAs. The minority carrier lifetime of AlGaAs is dominated by Shockley-Read-Hall recombination and exhibits a strong dependence on carrier density. A clear dependence on aluminum alloy composition is not observed, with minority carrier lifetimes ranging from nanoseconds to 10's of nanoseconds for $x < 0.4$ [14]. In the direct bandgap region for AlGaAs, the absorption coefficient exceeds 10^4 cm^{-1} at the band edge to approximately 5×10^5 cm^{-1} at the short wavelength edge of the visible region. Full optical absorption may therefore be obtained for a thickness of approximately 1 μm. For higher aluminum content where the band edge is near the cutoff of the indoor lighting spectrum, thicker layers are required for full absorption. For GaAs and lower aluminum content layers, thin layers of 0.5 μm or less are sufficient for full absorption.

Carrier collection in the base region for AlGaAs alloys can be a factor in determining conversion efficiency. The limiting mechanism for collection *via* minority carrier diffusion is hole transport in an n-type base. The hole mobility ranges from approximately 260 cm^2/V-s for GaAs to 80 cm^2/V-s at higher aluminum composition [15]. At the lower mobility value of 80 cm^2/V-s, and assuming a minority carrier lifetime of 1 ns, the minority carrier diffusion length is approximately 0.5 μm. This is near the edge of satisfying requirements for both high optical absorption and complete carrier collection.

The doping concentration of the base may be optimized to address a tradeoff between maintaining high carrier lifetime and mobility in the base while also maximizing the open-circuit voltage. The open-circuit voltage for an ideal one-sided junction diode with an n-type base may be represented by [16]

$$V_{OC} = \frac{kT}{q}\ln\left[\frac{(N_D + \Delta p)\Delta p}{n_i^2} + 1\right] \tag{9.2}$$

where kT/q is the thermal voltage, N_D is the donor concentration in the n-type base, n_i^2 is the intrinsic carrier concentration, and Δp is the excess carrier concentration due to illumination. Increasing the base doping concentration provides a higher open-circuit voltage up until doping concentrations approaching an order of magnitude of the emitter doping. However, increases in the doping concentration in the base may also lead to reduced minority carrier lifetime and mobility, resulting in reductions in photocurrent and non-ideal factors that will cause a decrease in open-circuit voltage in comparison to the ideal equation above. Typical doping concentrations for the base are on the order of 10^{17} cm^{-3}.

9.2.2 Contact

A top heavily-doped GaAs layer is typically incorporated to provide high-quality ohmic contacts. The heavily-doped layer is selectively removed outside of the metal contact layers, exposing the window layer and preventing optical absorption in this layer. For p-type layers, high doping concentrations near 10^{19} cm^{-3} may be achieved during the epitaxial growth process, providing a means to form low resistivity contacts with or without metal alloying with contact resistivity on the order of 10^{-6} Ω-cm^2 [17]. For n-type layers, alloyed contacts are often used due to limited doping levels achievable by epitaxial growth techniques, where contact resistivity $<10^{-6}$ Ω cm^2 is common [17]. For the photovoltaic cells in this work, the low contact resistivity associated with GaAs is easily sufficient to avoid degradation in PV cell performance. Series resistance issues will only occur at high illumination intensities, such as under concentrated sunlight [18].

9.2.3 Window

AlGaAs PV cells will typically utilize a top surface illumination geometry with a wide bandgap top window layer. The window layer serves as a transparent layer to couple light into the absorption region of the PV cell, while also providing an efficient means of reducing surface recombination. AlGaAs window layers cannot offer full transparency in the visible region, since the bandgap energy of AlAs at room temperature is approximately 2.2 eV. For $x > 0.45$, however, the bandgap energy is indirect, where transparency of thin layers will ultimately be limited by the larger direct bandgap. For AlAs at room temperature, this is approximately 3.14 eV [19]. This bandgap is still at the edge of the visible region, so full transparency is not achievable. Furthermore, some gallium content is typically used to avoid undesirable oxidation of the exposed surface. Semi-transparency throughout the visible indoor light spectrum may be achieved for AlGaAs windows with thickness of approximately 30 nm.

9.2.4 Emitter

For junction formation, a heavily-doped emitter typically utilizes the same AlGaAs composition as the base. Doping concentration is typically near degenerate doping, on the order of 10^{18} cm^{-3}. The choice of thickness presents a tradeoff between minimizing dark current and maximizing carrier collection. At the thin extreme, short diodes will result in an increase in reverse saturation current (dark current). Thick emitters may impede carrier collection due to reduced minority carrier diffusion lengths associated with the high doping concentration used in the emitter. Efficient carrier collection is critical in the emitter region since the photogeneration rate will be highest near the top illumination surface. An emitter thickness of approximately 200 nm provides excellent carrier collection while maintaining a low reverse saturation current.

9.2.5 Back Surface Field

A back surface field layer between the base and bottom contact layers is used to reduce surface recombination at this interface by reflecting minority carriers *via* an energetic barrier [20]. In the case of the PV cell structure shown in Table 9.1, the energetic barrier is achieved *via* an n-n^+ junction. Advantages of the back surface field become more important as the thickness of the base is reduced, which is common for optimized indoor PV cell designs in comparison to solar PV cell designs.

9.3 Large-Area AlGaAs III-V Photovoltaics

The relatively high cost of GaAs substrates (typically a factor of tens to hundreds higher than silicon) is prohibitive for direct use of photovoltaic cells on bulk substrates for large-area power generation. The cost of a 150 cm^2 GaAs photovoltaic cell is approximately \$200, whereas substrate cost is more than 80% of the total [21]. Epitaxial growth of III-V PV cell structures combined with wafer reuse can substantially reduce costs, where a techno-economic analysis suggests that a lower bound on cost per cell is achieved after approximately 20 reuse cycles with limiting factor associated with chemical mechanical polishing processes typically required for wafer planarization [21]. Efficient photovoltaic cells may be achieved for small layer thickness due to the strong optical absorption associated with AlGaAs and similar direct bandgap compound semiconductors. The optical absorption coefficients are on the order of 10^4 cm^{-1}, where full absorption of incident light may be achieved for layer thickness on the order of a few micrometers or less. Wafer reuse technologies incorporate a means of selectively removing the epitaxial device layer from the substrate, which include approaches such as spalling (mechanical removal) and epitaxial liftoff (ELO). The ELO technique involves selective removal of thin epitaxial structures from GaAs substrates through selective wet etching [22], and is among the most promising from a cost and technology standpoint. GaAs-based photovoltaic cells based on ELO have demonstrated efficiency comparable to cells on bulk GaAs substrates [23–27]. GaAs substrate reuse may be achieved when high-quality surfaces are obtained following sacrificial etches. Approaches for such wafer re-use include phosphide-based sacrificial layers [24] and the use of InGaP/GaAs protection layers adjacent to a sacrificial AlAs layer [26, 28]. The ELO process for the case of InGaP/GaAs protection layers is illustrated in Figure 9.3.

The thin-film ELO approach and capability of wafer re-use make III-V photovoltaics a possible candidate for indoor photovoltaics applications. An illustration of resulting ELO GaAs cells from Ref. [26] is shown in Figure 9.4, where a high cell conversion efficiency of >18% is achieved for a sequence of ELO steps on the same GaAs wafer. Under simulated AM1.5 sunlight, the record GaAs conversion efficiency in 2018 is 29.3% (Alta Devices) [29]. This record efficiency uses a thin-film cell based on ELO, which is higher than for bulk GaAs, by taking advantage of photon recycling [30]. The photon recycling approach provides an enhancement in conversion efficiency for photovoltaic cells that operate near the radiative limit, and is a result of restricting light emission from the cell.

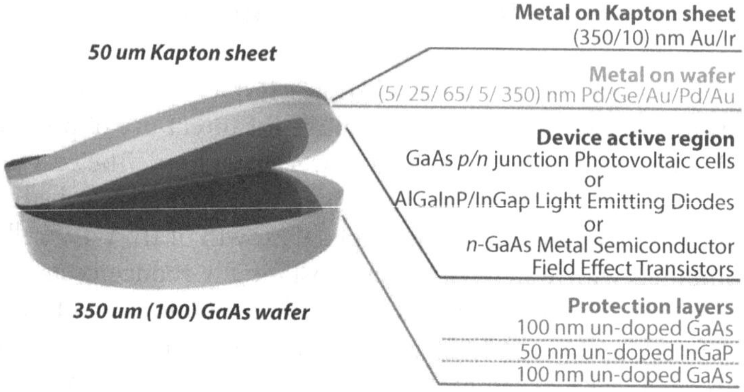

Figure 9.3 Schematic illustration of the generalized wafer structure used in nondestructive epitaxial lift-off (ND-ELO). The active thin film device region is lifted-off by selectively etching the AlAs sacrificial layer using dilute hydrofluoric acid. Details of the epitaxial layers, including the alternating GaAs-InGaP protection layers and the AlAs sacrificial layer, are shown along with the metal layer used in cold-welding to the plastic substrate. The active region structure is varied according to the application requirements, e.g., a photovoltaic cell, light-emitting diode, or metal semiconductor field effect transistor. (From Ref. [26])

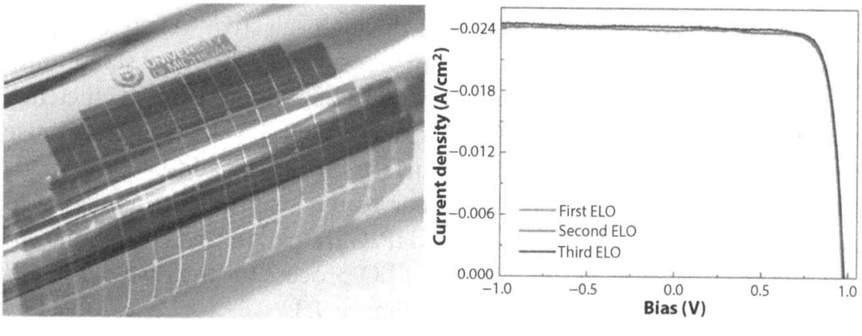

Figure 9.4 Close-up image of the GaAs thin film photovoltaic cell array and current density versus voltage (J–V) characteristics measured under 1 sun, AM1.5G simulated solar illumination. (From Ref. [26])

GaAs ELO photovoltaic cells were tested under indoor lighting conditions by Mathews *et al.* [31] using compact fluorescent and LED light sources. These commercial GaAs cells, along with GaInP cells, demonstrate substantially higher conversion efficiency than current a-Si and dye-sensitized solar cells used for indoor energy harvesting, with more than a factor of two increase for light levels in the 200–1000 lux range. A credit-card-sized (45 cm^2) cell provides approximately 4 mW of power under 1000 lux (similar to a bright office space), and is suitable for powering wireless sensor nodes in many applications. While larger scale GaAs PV energy harvesting has not been reported in the literature, arrays of GaAs ELO thin-film photovoltaic cells have been reported to provide mobile power for military applications. The module consists of a 10×3 array of 20 cm^2 single-junction GaAs cells, and is semirigid, lightweight, and provides the ability to recharge a high-capacity lithium-ion battery. The conversion efficiency of the module is approximately 19% (over cell area) under AM1.5G conditions. The high conversion efficiency of GaAs-based PV cells is clearly attractive for indoor photovoltaics applications and motivate further study, while ultimate application will depend on continued breakthroughs in cost reductions for this technology. While the ELO approach substantially reduces costs due to wafer reuse, the costs associated with III-V substrates are not fully eliminated due to substantial costs for preparing substrates for each reuse cycle [21].

A key factor in reducing costs is related to the epitaxial crystal growth technology. In typical epitaxial crystal growth technologies, such as molecular beam epitaxy (MBE) and metal-organic chemical vapor deposition (MOCVD), the source materials and precursors are expensive, are not efficiently used, and have very slow growth rates (in the range of μm/hour). All of these factors contribute to a relatively high cost for the epitaxial growth process that would limit feasibility of using III-V photovoltaics, even with wafer re-use enabled by ELO technologies. Recently, a dynamic hydride vapor phase epitaxy (D-HVPE) technology [32] has been developed as a new route to low-cost III-V epitaxial growth. Conventional HVPE growth processes enable high III-V growth rates, up to 300 μm/hour for GaAs [33]. The HVPE technique combines high quality material with high throughput and high precursor material utilization. A downside to HVPE has traditionally been the inability to realize heterointerfaces that are needed for high-performance optoelectronic devices, including photovoltaics. The D-HVPE technique uses a reactor design with multiple growth chambers that each have an independent steady-state deposition reaction for a particular material in an epitaxial layer design, with high-quality heterointerfaces [32, 34, 35]. The D-HVPE technology has demonstrated single-junction GaAs cells with solar energy conversion efficiency >25.3%, utilizing growth rates of approximately 60 μm/hour [32]. Analysis suggests that solar cell manufacturing cost may be reduced from the current level of around $50/W to below $0.50/W [36]. These cost reductions would rely heavily on both substantial improvements in substrate reuse and development and scale-up of a technology such as D-HVPE. Such cost reductions are imperative for III-V PV cells such as AlGaAs to be competitive for indoor applications. The high performance of AlGaAs cells under indoor conditions, coupled with activity in reducing costs associated with III-V optoelectronics, are expected to drive continued interest in AlGaAs indoor PV.

9.4 Small-Area AlGaAs Photovoltaics

Low-power electronic circuitry, including wirelessly interconnected sensor nodes, promises to be a transformational technology that can enable unsurpassed interconnectivity and a paradigm shift commonly referred to as the Internet of Things (IoT). These low-power systems require a source of energy, ideally from ambient sources. Power density for various energy harvesting sources are discussed in Chapter 2: Micro Energy Harvesting, Chapter 3: Introduction to IPV, and also summarized in Table 9.2 [37]. Ambient indoor lighting can provide sufficient energy for most of these applications, with a power density of approximately 1 μW/mm^2 under dim lighting conditions. The energy-autonomous operation of millimeter-scale sensors requires a power supply under active and standby operation of approximately 10 μW and 0.5 nW, respectively [38–41].

The functionality of these systems is highly dependent on power supply, to minimally achieve perpetual operation and to ideally achieve a desired number of operations for the system in a given time. A large fraction of active mode operations is typically devoted to wireless communications or specialized needs associated with sensor circuitry such as signal amplification and processing. The total power required P_{total} can be represented by

$$P_{total} = \frac{NP_{active}t_{active} + P_{standby}\left(t_{total} - Nt_{active}\right)}{t_{total}} \tag{9.3}$$

Table 9.2 Energy harvesting sources (adopted from [37]).

Power source	Power density
Photovoltaics (outdoors)	15,000 μW/cm^2
Photovoltaics (indoors)	10 μW/cm^2
Thermoelectric (5°C gradient)	40 μW/cm^2
Piezoelectric (shoe inserts)	330 μW/cm^3
Vibration (building)	300 μW/cm^3
Ambient radio frequency	<1 μW/cm^2

where N is the number of operations within a total time period t_{total}, t_{active} is the time for each active operation, P_{active} is the power consumption under active mode, and $P_{standby}$ is the power consumption in standby mode. Rearranging terms, the number of number of operations is given by

$$N = \left(\frac{P_{total} - P_{standby}}{P_{active} t_{active} - P_{standby} t_{active}} \right) t_{total} \tag{9.4}$$

Under typical conditions where $P_{active} >> P_{standby}$, and relating the total power to the power of the source and efficiency for energy harvesting, the number of operations is limited to

$$N \leq \left(\frac{\eta P_{source} - P_{standby}}{P_{active} t_{active}} \right) t_{total} \tag{9.5}$$

where P_{source} is the power of the source, η is the efficiency of energy conversion. The product $P_{active} t_{active}$ is typically given in terms of energy for each operation, and the value of N/t_{total} describes the frequency of operations achievable. The value for energy for each operation $P_{active} t_{active}$ has the largest variation, since this is highly dependent on the type of application (e.g., ranging from low power for a simple temperature logger to higher power for an application such as an imager). Furthermore, a particular system may require a series of different operations with different energy requirements and different frequency of execution. For example, a data logger may measure and record data periodically, and then wirelessly stream information for a series of data points at lower frequency. An example scenario for a mm-scale node is given in Table 9.3.

GaAs and AlGaAs-based PV cells are promising candidates for indoor energy harvesting to power small-scale electronics under low-flux indoor illumination conditions. In contrast to large area power generation, performance becomes a more critical parameter, while the overall cost of the PV cells becomes a smaller fraction of the overall system cost. To illustrate, assuming a cost of \$200 for a 150 cm^2 GaAs PV cell [21] would scale to approximately \$0.02 for a 1 mm^2 cell. The dimensional scaling for mm-scale systems is favorable for the use of high-performance PV cells, where the cost of the unit becomes reasonable in comparison to the overall system cost, and the more common metric of cost per unit area is less important. At these reduced device dimensions, however, there are more

Table 9.3 Example parameters for power requirements for perpetual operation of a mm-scale system, where the source power of 1 μW corresponds to typical indoor lighting conditions of approximately 600 lux in a 1 mm^2 area.

Parameter	**Value**
P_{source}	1 μW
$P_{standby}$	0.5 nW
$P_{active}t_{active}$	200 nJ
η	10%
n/t_{total}	<0.5 operations/s

stringent requirements on dark current, recombination, and shunt leakage current in order to achieve high conversion efficiency. Recombination losses result from the annihilation of photogenerated electron-hole pairs prior to carrier collection. The low-flux conditions characteristic of indoor lighting make recombination losses an important parameter in determining the power conversion efficiency of a photovoltaic cell. Recombination can occur in the bulk, surface, or perimeter of the device, where the surface and perimeter factors become more significant as cell area decreases. The influence of perimeter recombination on the GaAs-based diodes and solar cells has been reported previously [42–47]. The low current density associated with PV cells under indoor illumination conditions typically makes losses due to series resistance insignificant. The reduced current density, however, requires higher shunt resistance to maintain high power conversion efficiency. In this section, we examine the performance of mm-scale AlGaAs PV cells under indoor illumination conditions.

9.4.1 Model of *J*-*V* Characteristics

The current-voltage (*J*-*V*) characteristics of PV cells, incorporating recombination losses and shunt resistance, are described by the following models. The current density is given by

$$J = J_{01}\,(e^{qV/kT}-1) + J_{02}\,(e^{qV/2kT}-1) \tag{9.6}$$

where J is total current density, q is the charge of an electron, k is the Boltzmann constant, T is temperature, and J_{01} and J_{02} represent reverse saturation current densities for recombination in the quasi-neutral region and recombination in the space charge region, respectively. In the case of GaAs and AlGaAs PV cells, the second term related to recombination in the space charge region is dominant. The bulk and perimeter components of the reverse saturation current density can be represented by

$$J_{02} = J_{02B} + J'_{02P}\left(\frac{P}{A}\right) \tag{9.7}$$

where J_{02B} is the bulk recombination current density and J'_{02P} is the perimeter recombination coefficient. The perimeter recombination coefficient can be written as

$$J'_{02P} = qn_i s_0 L_s \tag{9.8}$$

where n_i is the intrinsic carrier concentration, s_0 is the surface recombination velocity, and L_s is the effective surface diffusion length. The product s_0L_s for high-efficiency GaAs solar cells have been reported as less than 1 cm²/s [45, 46], where passivation of surfaces and sidewalls are critical for small area cells.

Parasitic shunt conductance can degrade cell performance, and becomes increasingly important under low-flux conditions where short-circuit current values are small. The parasitic resistance components for a photovoltaic cell are described by

$$J = J_{SC} - J_0\left[\exp\left(\frac{q(V+JAR_s)}{nkT}\right)-1\right]-\frac{V+JAR_S}{AR_{sh}} \tag{9.9}$$

where J_{sc} is the short-circuit current density, and R_s and R_{sh} are series and shunt resistance, respectively. Note that in this case, the reverse leakage current is represented by J_0 and ideality factor n for the range of cases where current is limited by recombination in the bulk region or space charge region. Shunt resistance values may be obtained from current-voltage measurements under extremely dim illumination, where J_{sc} and V_{oc} exhibit a near linear relationship when shunt conductance is dominant. This technique can be used to directly measure the shunt resistance of a solar cell without prior knowledge of solar cell parameters [48]. Variable light intensity measurements where shunt resistance is dominant allow for this determination, without need for accurate knowledge of the source intensity. Neglecting series resistance, J_{sc} and V_{oc} may be related by

$$J_{SC} - J_0\left[\exp\left(\frac{qV_{oc}}{nkT}\right) - 1\right] = \frac{V_{oc}}{AR_{sh}}. \tag{9.10}$$

Under dim conditions, the J_{sc} term becomes dominant, resulting in the linear relationship

$$J_{SC} = \frac{V_{oc}}{AR_{sh}}. \tag{9.11}$$

To illustrate the impact of shunt resistance on diodes under typical indoor lighting conditions, simulated J-V and P-V curves are shown in Figure 9.5 for a diode (not material specific) with a reverse current density

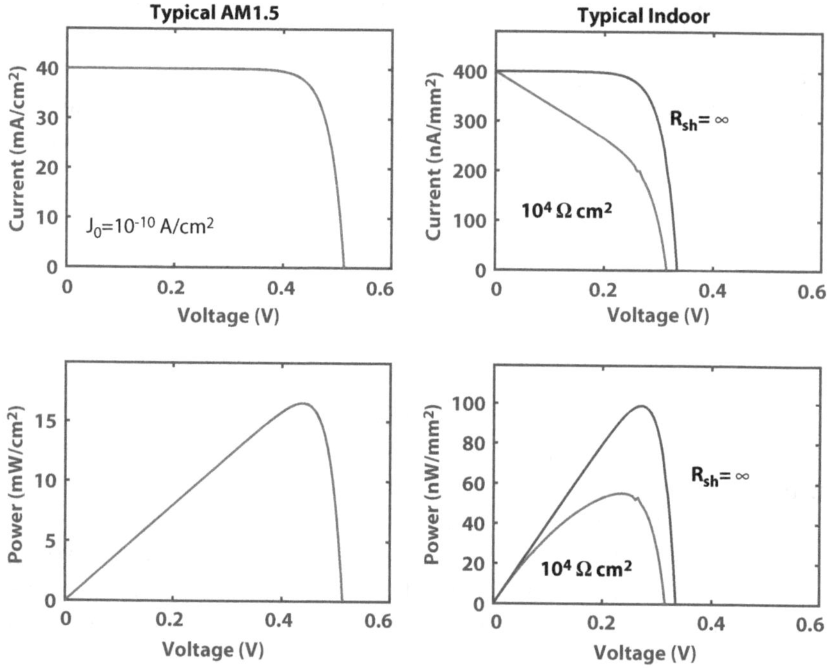

Figure 9.5 Simulated J-V and P-V curves for PV cells under typical AM1.5 and indoor lighting conditions assuming a reverse saturation current density of $J_0 = 10^{-10}$ A/cm^2.

of 10^{-10} A/cm^2, ideality factor of $n = 1$, and shunt resistance values of infinity (ideal) and 10^4 Ω-cm^2. A value of $J_L = 40$ mA/cm^2 is assumed under AM1.5 conditions, and a factor of 1,000 smaller for indoor illumination conditions (400 nA/mm^2). The solar cell *J-V* and *P-V* results under AM1.5 conditions are relatively typical of a silicon PV cell, with no observable difference for cases of 10^4 Ω-cm^2 and infinite shunt resistance. Under the much weaker illumination conditions of indoor lighting, the open-circuit voltage drops significantly, and the finite shunt resistance dramatically degrades fill factor and the maximum power point.

9.4.2 Performance of mm-Scale AlGaAs Photovoltaics

Example characteristics of an AlGaAs PV cell that is optimized for indoor conditions are shown in Figure 9.6. As described in this chapter previously, the primary differences from a conventional GaAs solar cell are the wider bandgap for the base region ($Al_{0.2}Ga_{0.8}As$) in comparison to GaAs and the reduced thickness of the base region (less than 1 μm for complete absorption). A comparison of the external quantum efficiency of PV cells with GaAs and AlGaAs absorbers is shown in Figure 9.7, and compared to the light output spectrum of a white phosphor LED typical of many indoor

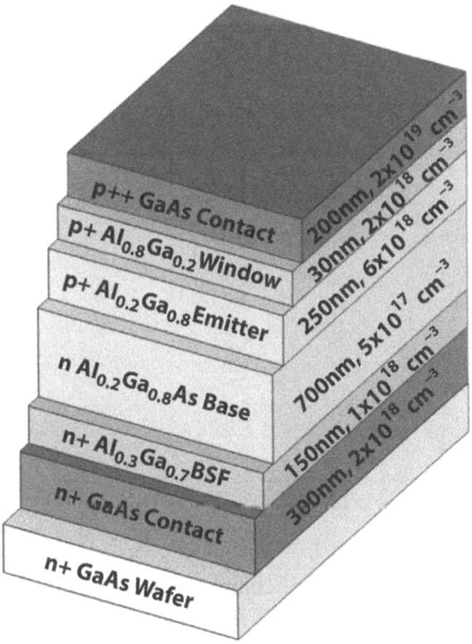

Figure 9.6 Schematic diagram of an AlGaAs PV cell optimized for indoor lighting conditions.

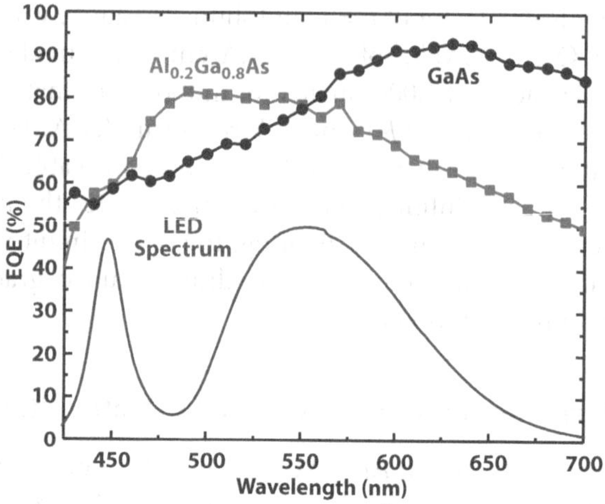

Figure 9.7 External quantum efficiency comparison of PV cells with GaAs and AlGaAs base regions and comparison to an example spectrum of a white phosphor LED.

lighting conditions. The external quantum efficiency of the PV cell with $Al_{0.2}Ga_{0.8}As$ absorber provides a more appropriate match to the indoor lighting spectrum.

The *J-V* and *P-V* characteristics comparing PV cells with GaAs and $Al_{0.2}Ga_{0.8}As$ base regions under white LED illumination are shown in Figure 9.8 for cells with area of 1.22 mm². As predicted, the larger bandgap

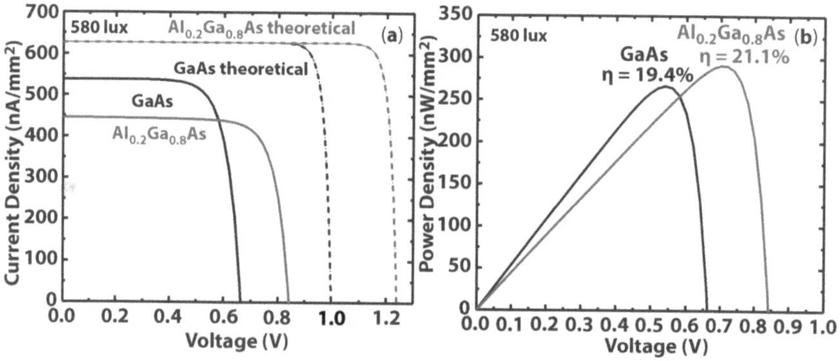

Figure 9.8 Measured (a) J-V and (b) P-V characteristics comparing GaAs and AlGaAs PV cells under white light illumination with cell area of 1.22 mm². Comparison is shown to theoretical values based on the energy bandgap of the PV cell base material. White LED illumination used a Vollong 5W high power white LED VL-H01xWx5 and lux was measured using an Extech LT300 light meter.

energy of the $Al_{0.2}Ga_{0.8}As$ PV cell results in higher open-circuit voltage and higher voltage for the maximum power point. The overall efficiency at the maximum power point of the $Al_{0.2}Ga_{0.8}As$ PV cell is also increased, where 21.1% is observed in comparison to 19.4% for GaAs. While the $Al_{0.2}Ga_{0.8}As$ PV cells show increased voltage and power conversion efficiency, there are still greater losses in comparison to the GaAs PV cells, as observed in the reduced short-circuit current density. The reduced short-circuit current density is consistent with the reduced external quantum efficiency shown in Figure 9.7. This may reflect a combination of factors, including incomplete optical absorption at longer wavelengths, reduced carrier collection due to shorter minority carrier diffusion length in the AlGaAs base and emitter, and increased non-radiative losses in higher aluminum content AlGaAs alloys.

The dependence of the GaAs and $Al_{0.2}Ga_{0.8}As$ PV cells on light intensity was investigated and compared to commercial a-Si and crystalline silicon cells, as shown in Figure 9.9. The a-Si cells exhibit a very stable intensity dependence, though the overall power conversion efficiency is limited to below 10% for the conditions tested. In comparison to performance under AM1.5 conditions, the crystalline silicon cell shows a dramatic reduction in power conversion efficiency under the indoor illumination conditions, with efficiency lower than the a-Si cells and a strong intensity dependence.

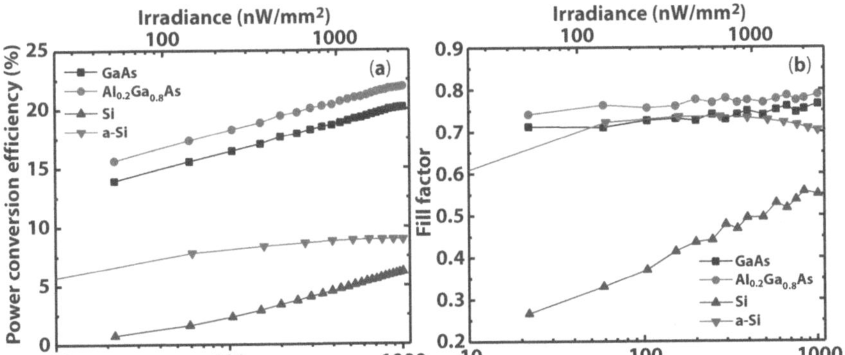

Figure 9.9 Power conversion efficiency and fill factor versus white LED illumination intensity for GaAs and AlGaAs PV cells with cell area of 1.22 mm². Performance is compared to commercial a-Si (CYMBET CBC-PV-02, 1797 mm² area) and crystalline silicon (IXOLAR KXOB22-12X1, 22% efficiency at AM1.5, 154 mm² area). White LED illumination used a Vollong 5W high power white LED VL-H01xWx5 and lux was measured using an Extech LT300 light meter.

The reduced conversion efficiency is apparent in a highly degraded fill factor, and related to shunt resistance, as described in the simulations of Figure 9.5. The highly degraded efficiency of crystalline silicon solar cells under indoor lighting conditions is consistent with prior reports [1, 4]. The GaAs and AlGaAs PV cells provide dramatically higher conversion efficiency, and provide a relatively stable intensity dependence. From these results, it is clear that the shunt resistance of GaAs and AlGaAs PV cells are sufficient to maintain high conversion efficiency for a range of typical indoor lighting conditions. Measurements of the GaAs cells under low flux conditions in the regime where the approximation $J_{sc} = V_{oc}/AR_{sh}$ is valid resulted in shunt resistance of greater than 5 MΩ-cm² [49]. This value is sufficiently large to preserve PV cell fill factor, which is estimated to begin degrading at approximately 0.1 MΩ-cm² and lower [49].

9.4.3 Dark Current Limitations

Efficiency limitations for the GaAs and AlGaAs cells, and change from the theoretical limit for efficiency shown in Figure 9.8, are related to the diode dark current (reverse saturation current under dark conditions). Dark current for an example GaAs PV cell is shown in Figure 9.10, where there is a clear exponential behavior in the forward bias region until approximately 0.8 V. At higher voltages, a series-resistance limited behavior is observed. Numerical fitting of the exponential region of forward bias results in diodes

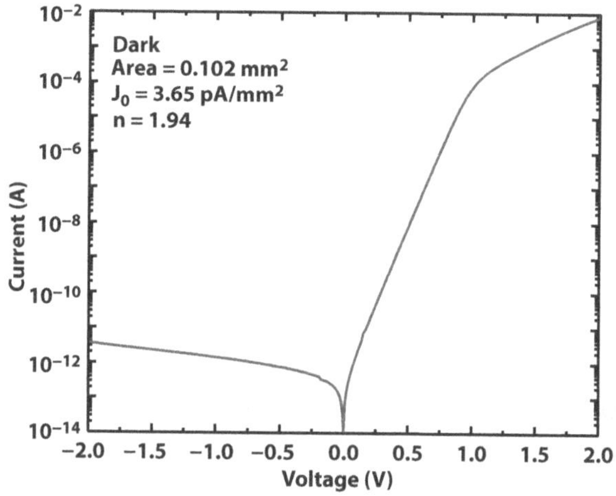

Figure 9.10 Dark current density of GaAs PV cell with area of 0.172 mm² and extracted value of reverse saturation current and ideality factor in the exponential forward bias region.

with an ideality factor near $n = 2$, with values for the reverse leakage current J_0, shows a strong perimeter/area (P/A) dependence for cells with area of less than 1 mm^2. The P/A dependence of dark current density of these GaAs PV cells is show in Figure 9.11. Extracted values for bulk and perimeter components of dark current are $J_{02B} = 140$ fA/mm^2 and $J'_{02P} = 170$ fA/mm, and are similar to prior reports for high efficiency GaAs solar cells [45, 46]. Relating the perimeter dark current to surface/sidewall recombination, the product s_0L_s for these devices is determined to be 5.1 cm^2/s, which is somewhat higher than the value of 1 cm^2/s reported for high-efficiency GaAs solar cells. These particular PV cells were passivated using $(NH_4)_2S$ surface treatment and Si_3N_4 passivation using the procedure reported in Ref. [50]. Further studies to passivate perimeter recombination will be critical in improving the performance of GaAs-based photovoltaic for low-flux and small-area applications.

The measured dark current values provide a useful means of projecting the dependence of power conversion efficiency on illumination intensity and PV cell dimension. Calculated values of power conversion efficiency assuming the measured dark current values for GaAs PV cells in this work are shown in Figure 9.12. The values for side length assume a PV cell with a square geometry. The PV cell efficiency begins to reach the maximum value for a 1 cm^2 cell, which is limited by the bulk recombination mechanism described by J_{02B}. The power conversion efficiency shows a strong dependence on PV cell size when the size decreases to approximately 1 mm^2 and

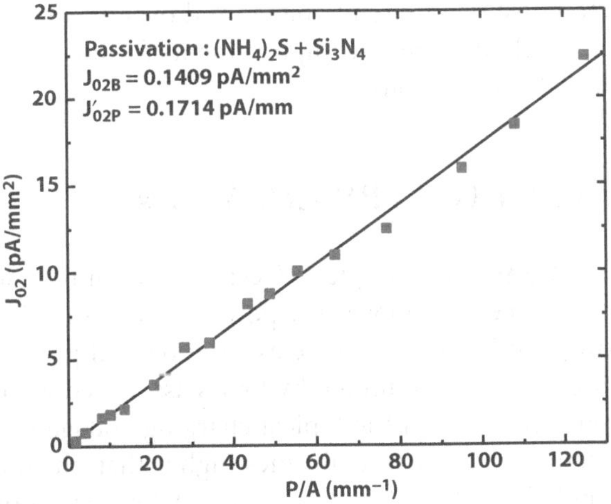

Figure 9.11 Reverse saturation current density of dominant space charge region recombination component versus perimeter/area for GaAs cells with variable area.

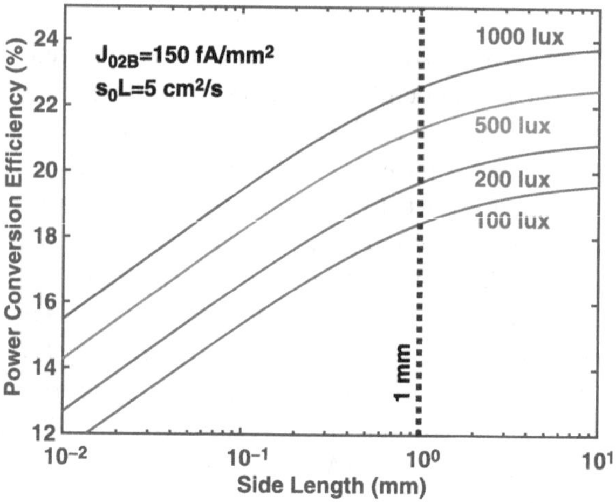

Figure 9.12 Calculated power conversion efficiency versus side length for PV cells at varying indoor illumination conditions assuming measured values for bulk and perimeter dark current. Illumination assumed a lux to photon ratio of 0.125 μmol/m^2/s.

below, limited by perimeter recombination described by J'_{02P}. This result highlights the importance of perimeter recombination for small-area cells under low-flux conditions, which may often be neglected for larger cells under typical solar irradiation. Chemical studies have shown improvements in the passivation of GaAs surface recombination using sulfides and thiols [51–54]. While such techniques have typically been employed in solar cells to boost open-circuit voltage and overall efficiency, they can similarly offer a path for dramatic improvements for small-area cells operating under low-flux conditions.

9.5 Monolithic GaAs PV Cell Arrays

High-efficiency GaAs-based single PV cells at the mm-scale show outstanding performance with above 20% power conversion efficiency under indoor lighting condition. Nevertheless, the overall power generation from these single PV cells is limited by losses associated with voltage up-conversion required to reach the typical charging voltage of batteries for mm-scale systems, which is several times higher than the output voltage (<1 V) of a single PV cell [31]. Typically, a switched capacitor network is used for voltage up-conversion, where resistive and switching losses limit the efficiency to 50% [55]. A promising alternative is to direct connect PV cells in series, where the power conversion efficiency above 80% has been demonstrated through bypassing voltage up-conversion [55]. However, wafer-scale monolithic PV modules with series connected diodes are very challenging due to requirements to electrically isolate individually diode devices, and to avoid shunt leakage paths through the shared substrate [56]. Monolithic silicon PV modules present several challenges due to relatively low open-circuit voltage and low absorption strength, requiring a larger number of cells (more susceptible to shunt leakage degradation) and thicker absorber regions (problematic for device isolation and interconnection). In this manner, GaAs or other III-V-based PV arrays are more attractive solutions for monolithic PV modules due to the larger voltage generation (minimizing the number of series connections) and higher absorption strength (ability to use very thin device layers).

An example of a monolithic PV array on a semi-insulating substrate is shown in Figure 9.13, along with a schematic equivalent circuit diagram illustrating shunt leakage. It should be noted that this shunt leakage between series connected diodes differs from the shunt leakage path across the junction described in the previous section. The high-resistivity of semi-insulating substrates such as GaAs is typically sufficient for electrical device isolation for many applications, but is not necessarily sufficient for monolithic PV cells. A blocking junction diode may be incorporated between PV cells and the substrate to provide an additional measure for electrical isolation. A further consideration of importance when assessing shunt leakage current paths is the possibility of light-activated shunt leakage, where illumination conditions may increase leakage current due to photoconductivity of the semi-insulating substrate material, or photoresponse of the blocking diode.

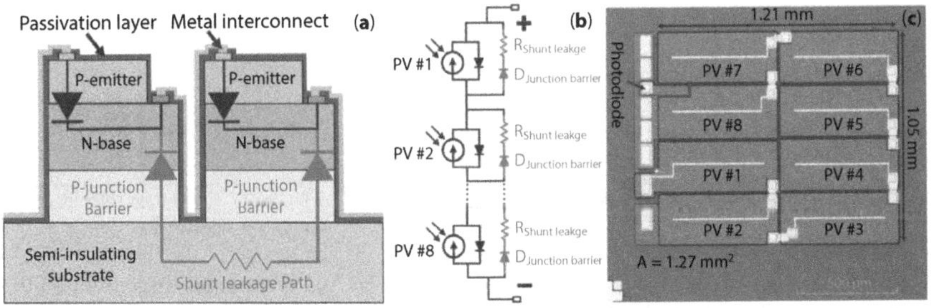

Figure 9.13 Schematic diagrams of (a) device structure illustrating shunt leakage path and incorporated junction barrier, and (b) equivalent circuit model of the PV module. (c) Optical microscope image of fabricated PV module.

Results from monolithic PV arrays of GaAs cells will be described in the following [44]. Three approaches are investigated to characterize the shunt leakage current and possible degradation of device performance: 1) use of semi-insulating substrate alone, 2) incorporation of a p-GaAs diode junction barrier, and 3) incorporation of a p-$Al_{0.3}Ga_{0.7}As$ diode junction barrier. The fabricated PV module has a series connection of 8 single PV cells with a total module area of 1.27 mm^2, as shown in Figure 9.13. PV cells are isolated with 10 μm trenches between adjacent devices. The expected voltage output for the 8-cell series connection is approximately 5 V under typical indoor lighting conditions, with the intent to provide sufficient voltage for direct battery charging.

The current-voltage (*J-V*) and power-voltage (*P-V*) characteristics of PV modules with p-GaAs and p-$Al_{0.3}Ga_{0.7}As$ junction barrier isolation are shown in Figure 9.14 for white LED illumination at 586 lux (1.4 μW/mm^2). Monolithic PV modules with both p-GaAs and p-AlGaAs barrier isolation produce an open-circuit voltage around 5 V, which is sufficient for direct battery charging without voltage up-conversion as designed. However, the PV module with p-GaAs barrier demonstrates a drastic degradation in fill factor from 0.752 to 0.565 in comparison to calculated *J-V* characteristic for no shunt leakage current. The corresponding power conversion efficiency under white LED illumination decreases from 17.7% to 10.97% due to insufficient electrical isolation between cells by the GaAs junction barrier under illumination conditions. The incorporation of p-AlGaAs barrier

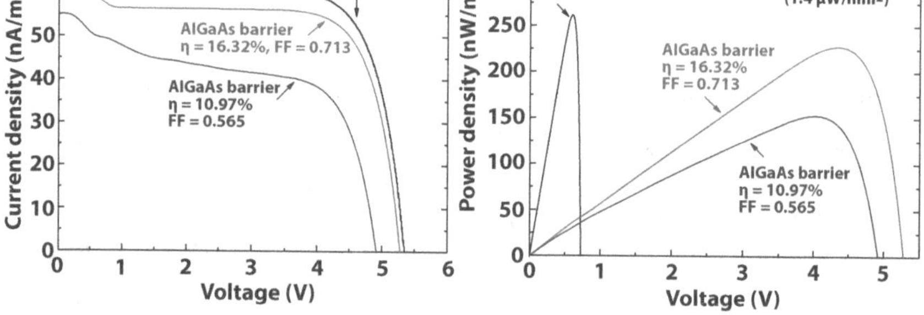

Figure 9.14 Measured J-V and P-V characteristics of GaAs PV modules under white LED illumination with GaAs and AlGaAs junction barrier isolation. Comparisons are shown to simulated results (dashed) without shunt leakage current and to the power density of a single PV cell. White LED illumination used a Vollong 5W high power white LED VL-H01xWx5 and lux was measured using an Extech LT300 light meter.

demonstrates the dramatic improvement in device performance, where the measured overall power conversion efficiency of 16.32% is comparable to the calculated value of 17.7% without the shunt leakage. Overall, there is an obvious shunt leakage dependence on device performance for the monolithic PV array, requiring detailed characterization of shunt leakage current in order to understand limiting factors for monolithic PV arrays.

The shunt leakage current between bottom n-contacts of adjacent PV cells without series metal interconnection between adjacent diodes is shown in Figure 9.15. The approximately linear current-voltage characteristic of shunt leakage through the semi-insulating substrate without junction barrier isolation suggests that shunt leakage is highly dependent on the resistivity of the semi-insulating substrate. A clear reduction in leakage current is observed when incorporating p-GaAs and p-$Al_{0.3}Ga_{0.7}As$ blocking junction diodes. The shunt leakage current for the p-GaAs junction barrier sample shows a near exponential dependence, suggesting that the leakage is limited by the energy barrier height of p-type junction barrier. A further reduction in shunt leakage is observed when incorporating the p-AlGaAs junction barrier, where the current is reduced to the pA level. The further reduction in shunt leakage current is attributed to the increased energy barrier height provided by the wider bandgap AlGaAs material. While the three approaches show varying degrees of leakage current, all three device isolation schemes demonstrate shunt leakage current near or below the nA range, which would not be expected to degrade the device performance dramatically. This contradicts the observed degradation in *J-V* characteristics of the modules shown in Figure 9.14. Therefore,

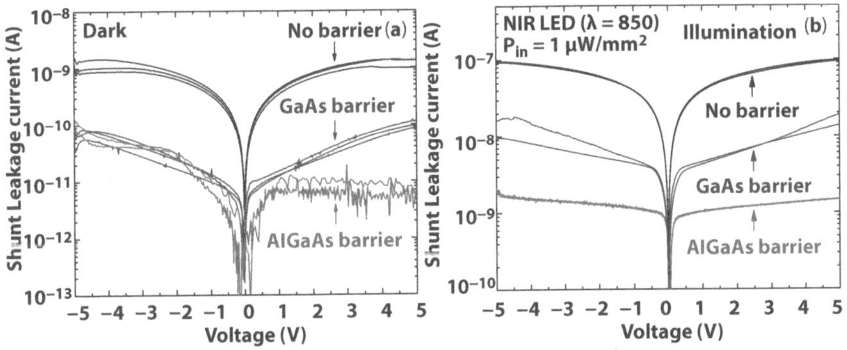

Figure 9.15 Measured shunt leakage current (a) under dark condition and (b) under illumination for three different barrier structures: no barrier, p-GaAs junction, and p-$Al_{0.3}Ga_{0.7}As$ junction.

the shunt leakage current characteristics under dark conditions are not sufficient in explaining the behavior of PV modules under illumination.

The measured shunt leakage current under near-infrared (NIR, 850 nm) illumination shows that there is a substantial increase in the shunt leakage current by a factor of approximately 100 under illumination. This undesirable photoactivated behavior is attributed to an increase in photoconductivity at the junction barrier and/or exposed regions of the semi-insulating substrate [13]. The shunt leakage for p-GaAs barrier isolation increases the level of 10 nA, comparable to photogenerated current of PV module under low-flux indoor light conditions. In contrast, the increase in the energy barrier height by the p-AlGaAs barrier blocks shunt leakage down to the level of 1 nA under illumination, which preserves the fill factor and the overall power conversion efficiency of the PV array.

In comparison to the 6.4 mm^2 single PV cell shown in Figure 9.14, the PV array with AlGaAs barrier isolation produces eight times higher voltage around 5 V, with 1/8 reduction in current. The generated voltage is sufficient for direct battery charging directly and the PV module provides high-efficiency voltage up-conversion process of approximately 90% under white LED illumination. The conversion efficiency and fill factor for modules under variable incident light intensity are shown in Figure 9.16. The results demonstrate that the p-AlGaAs blocking diode isolation preserves the device performance to be comparable to a single PV across a broad range of illumination intensity, including extremely dim light illumination

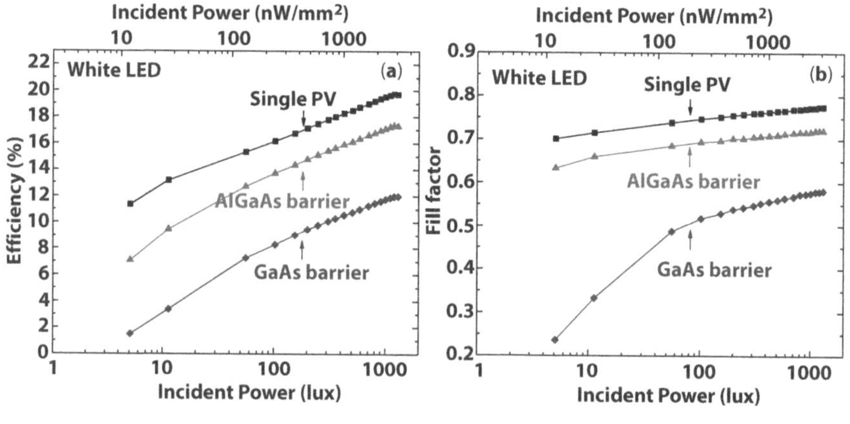

Figure 9.16 Comparison of measured conversion efficiency and fill factor under white LED illumination at variable power. White LED illumination used a Vollong 5W high power white LED VL-H01xWx5 and lux was measured using an Extech LT300 light meter.

below 100 lux. The degraded conditions of the PV modules with GaAs blocking diodes emphasize the critical importance of limiting shunt leakage current to realize monolithic PV arrays with high energy conversion efficiency.

9.6 Conclusion

GaAs and AlGaAs PV cells offer high power conversion efficiency under typical indoor lighting conditions. Wafer reuse techniques such as ELO, coupled with low-cost epitaxial techniques such as D-HVPE, provide an opportunity for a viable indoor PV technology that is feasible in cost. However, substantial work remains in testing these cells under indoor conditions, optimizing device structures for the indoor spectrum and light intensity, and in the integration of these cells into indoor PV systems. Small-area GaAs and AlGaAs PV have been characterized in more detail under indoor illumination conditions with the objective of achieving self-powered mm-scale systems for IoT applications. PV devices designed for indoor conditions demonstrate power conversion efficiency exceeding 20%, where the use of AlGaAs alloys in the PV cell base provides an increase in operating voltage and higher efficiency relative to a comparable cell with GaAs base due to the increase in bandgap energy. The power density provided by these cells under typical indoor lighting conditions of approximately 500 lux is sufficient in meeting the demands for perpetual operation of present-day mm-scale low-power systems. While the power conversion efficiency of these cells is among the highest to date under indoor conditions, substantial gains are possible to achieve the theoretical limiting efficiency. The GaAs-based devices on the mm-scale and below are found to be limited by perimeter recombination, where further effort to passivate exposed material sidewalls is expected to improve conversion efficiency. Efficiency gains for indoor lighting may also be obtained through a more thorough design optimization of cell design and AlGaAs base layer composition. Energy harvesting for mm-scale IoT applications may also be advanced using monolithic modules of series-connected PV cells. These modules can provide efficient voltage up-conversion capable of direct battery charging without the need for lossy voltage up-conversion circuitry. Efficient indoor PV power conversion may be achieved with AlGaAs devices, where continued exploration of improved designs, mitigation of non-idealities, and integration into application spaces are expected to result in a robust energy harvesting solution.

References

1. Freunek, M., Freunek, M., Reindl, L.M., Maximum efficiencies of indoor photovoltaic devices. *IEEE J. Photovoltaics*, 3, 59–64, 2013.
2. Shockley, W. and Queisser, H.J., Detailed balance limit of efficiency of p-n junction solar cells. *J. Appl. Phys.*, 32, 510–519, 1961.
3. Randall, J.F. and Jacot, J., Is AM1.5 applicable in practice? Modelling eight photovoltaic materials with respect to light intensity and two spectra. *Renewable Energy*, 28, 1851–1864, 2003.
4. Reich, N.H., van Sark, W. G. J. H. M., Alsema, E.A., Lof, R.W., Schropp, R.E.I., Sinke, W.C., Turkenburg, W.C., Crystalline silicon cell performance at low light intensities. *Sol. Energy Mater. Sol. Cells*, 93, 1471–1481, 2009.
5. Bunea, G.E., Wilson, K.E., Meydbray, Y., Campbell, M.P., Ceuster, D.M.D., Low light performance of mono-crystalline silicon solar cells, in: *2006 IEEE 4th World Conference on Photovoltaic Energy Conference*, vol. 2, pp. 1312–1314, Waikoloa, HI, 2006.
6. Beeby, S. and White, N., *Energy Harvesting for Autonomous Systems*, Artech House, London, 2010.
7. Ohno, Y., Spectral design considerations for white LED color rendering. *Opt. Eng.*, 44, 11, 111302, 2005.
8. GaAs provides the key ingredient for 5G phones. *Compd. Semicond.*, 23, 8, 14, 2017.
9. Virshup, G.F., Ford, C.W., Werthen, J.G., A 19% efficient AlGaAs solar cell with graded band gap. *Appl. Phys. Lett.*, 47, 1319–1321, 1985.
10. Melloch, M.R., Tobin, S.P., Bajgar, C., Keshavarzi, A., Stellwag, T.B., Lush, G.B., Lundstrom, M.S., Emery, K., High-efficiency Al0.22Ga0.78As solar cells grown by molecular beam epitaxy. *Appl. Phys. Lett.*, 57, 52–54, 1990.
11. Amano, C., Sugiura, H., Ando, K., Yamaguchi, M., Saletes, A., High-efficiency Al0.3Ga0.7As solar cells grown by molecular beam epitaxy. *Appl. Phys. Lett.*, 51, 1075–1077, 1987.
12. Ragay, F.W., Martí, A., Araújo, G.L., Wolter, J.H., Experimental analysis of the efficiency of heterostructure GaAs-AlGaAs solar cells. *Sol. Energy Mater. Sol. Cells*, 40, 5–21, 1996.
13. Datas, A., López, E., Ramiro, I., Antolín, E., Martí, A., Luque, A., Tamaki, R., Shoji, Y., Sogabe, T., Okada, Y., Intermediate band solar cell with extreme broadband spectrum quantum efficiency. *Phys. Rev. Lett.*, 114, 157701, 2015.
14. Ahrenkiel, R.K., Chapter 2 Minority-Carrier Lifetime in III–V Semiconductors, in: *Semiconductors and Semimetals*. R. K. Ahrenkiel and M. S. Lundstrom (eds.), pp. 39–150, Elsevier, 1993.
15. Look, D.C., Lorance, D.K., Sizelove, J.R., Stutz, C.E., Evans, K.R., Whitson, D.W., Alloy scattering in p-type AlxGa1–xAs. *J. Appl. Phys.*, 71, 260–266, 1992.
16. Sinton, R.A. and Cuevas, A., Contactless determination of current–voltage characteristics and minority-carrier lifetimes in semiconductors from quasi-steady-state photoconductance data. *Appl. Phys. Lett.*, 69, 2510–2512, 1996.

17. Baca, A.G., Ren, F., Zolper, J.C., Briggs, R.D., Pearton, S.J., A survey of ohmic contacts to III-V compound semiconductors. *Thin Solid Films*, 308–309, 599–606, 1997.
18. Yoshida, S., Mitsui, K., Oda, T., Yoshida, M., Yukimoto, Y., Series resistance effects and cell performances in GAAS concentrator solar cells. *Electr. Eng. Jpn.*, 103, 122–128, 1983.
19. Yim, W.M., Direct and indirect optical energy gaps of AlAs. *J. Appl. Phys.*, 42, 2854–2856, 1971.
20. v. Roos, O., A simple theory of back surface field (BSF) solar cells. *J. Appl. Phys.*, 49, 3503–3511, 1978.
21. Ward, J.S., Remo, T., Horowitz, K., Woodhouse, M., Sopori, B., VanSant, K., Basore, P., Techno-economic analysis of three different substrate removal and reuse strategies for III-V solar cells. *Prog. Photovoltaics Res. Appl.*, 24, 1284–1292, 2016.
22. Konagai, M., Sugimoto, M., Takahashi, K., High efficiency GaAs thin film solar cells by peeled film technology, in: *Proceedings of the Fourth International Conference on Vapour Growth and Epitaxy*, Netherlands, 9–13 July 1978, 1978.
23. Kayes, B.M., Hui, N., Twist, R., Spruytte, S.G., Reinhardt, F., Kizilyalli, I.C., Higashi, G.S., 27.6% Conversion efficiency, a new record for single-junction solar cells under 1 sun illumination, in: *2011 37th IEEE Photovoltaic Specialists Conference (PVSC 2011)*, Piscataway, NJ, USA, 19–24 June 2011, 2011.
24. Cheng, C.-W., Shiu, K.-T., Li, N., Han, S.-J., Shi, L., Sadana, D.K., Epitaxial lift-off process for gallium arsenide substrate reuse and flexible electronics. *Nat. Commun.*, 4, 1577, 2013.
25. Kayes, B.M., Ling, Z., Twist, R., Ding, I.K., Higashi, G.S., Flexible thin-film tandem solar cells with 30% efficiency. *IEEE J. Photovoltaics*, 4, 729–33, 2014.
26. Lee, K., Zimmerman, J.D., Hughes, T.W., Forrest, S.R., Non-destructive wafer recycling for low-cost thin-film flexible optoelectronics. *Adv. Funct. Mater.*, 24, 4284–4291, 2014.
27. Tatavarti, R., Hillier, G., Dzankovic, A., Martin, G., Tuminello, F., Navaratnarajah, R., Du, G., Vu, D.P., Pan, N., Lightweight, low cost GaAs solar cells on 4 epitaxial liftoff (ELO) wafers, in: *33rd IEEE Photovoltaic Specialists Conference, PVSC 2008*, San Diego, CA, United states, May 11, 2008 - May 16, 2008, 2008.
28. Lee, K., Zimmerman, J.D., Xiao, X., Sun, K., Forrest, S.R., Reuse of GaAs substrates for epitaxial lift-off by employing protection layers. *J. Appl. Phys.*, 111, 033527, 2012.
29. Tablero, C., Analysis of the electronic properties of intermediate band materials as a function of impurity concentration. *Phys. Rev. B: Condens. Matter Mater. Phys.*, 72, 35213–1, 2005.
30. Kosten, E.D., Kayes, B.M., Atwater, H.A., Experimental demonstration of enhanced photon recycling in angle-restricted GaAs solar cells. *Energy Environ. Sci.*, 7, 1907–1912, 2014.

31. Mathews, I., King, P.J., Stafford, F., Frizzell, R., Performance of III–V solar cells as indoor light energy harvesters. *IEEE J. Photovoltaics*, 6, 230–235, 2016.
32. Simon, J., Schulte, K.L., Horowitz, K.A.W., Remo, T., Young, D.L., Ptak, A.J., III-V-based optoelectronics with low-cost dynamic hydride vapor phase epitaxy. *Crystals*, 9, 3, 2018.
33. Deschler, M., Gruter, K., Schlegel, A., Beccard, R., Jurgenson, H., Balk, P., Very rapid growth of high quality GaAs, InP, and related III-V compounds. *J. Phys. Colloques*, 49, C4-689–C4-692, 1988.
34. Jain, N., Simon, J., Schulte, K.L., Friedman, D.J., Diercks, D.R., Packard, C.E., Young, D.L., Ptak, A.J., Tunable bandgap GaInAsP solar cells with 18.7% photoconversion efficiency synthesized by low-cost and high-growth rate hydride vapor phase epitaxy. *IEEE J. Photovoltaics*, 8, 1577–1583, 2018.
35. Schulte, K.L., Simon, J., Mangum, J., Packard, C.E., Gorman, B.P., Jain, N., Ptak, A.J., Development of GaInP solar cells grown by hydride vapor phase epitaxy. *IEEE J. Photovoltaics*, 7, 1153–1158, 2017.
36. Horowitz, K.A.W., Remo, T., Smith, B., Ptak, A., Techno-economic analysis and cost reduction roadmap for III-V solar cells. *Nat. Renewable Energy Lab. Tech. Rep.*, 2018. NREL/TP-6A20-72103.
37. Roundy, S., Wright, P.K., Rabaey, J.M., *Energy Scavenging for Wireless Sensor Networks*, Springer, Boston, MA, 2004.
38. Lee, Y., Bang, S., Lee, I., Kim, Y., Kim, G., Ghaed, M.H., Pannuto, P., Dutta, P., Sylvester, D., Blaauw, D., A modular 1 mm^3 die-stacked sensing platform with low power I^2C inter-die communication and multi-modal energy harvesting. *IEEE J. Solid-State Circuits*, 48, 229–243, 2013.
39. Jung, W., Oh, S., Bang, S., Lee, Y., Foo, Z., Kim, G., Zhang, Y., Sylvester, D., Blaauw, D., An ultra-low power fully integrated energy harvester based on self-oscillating switched-capacitor voltage doubler. *IEEE J. Solid-State Circuits*, 49, 2800–2811, 2014.
40. Hanson, S., Seok, M., Lin, Y., Foo, Z., Kim, D., Lee, Y., Liu, N., Sylvester, D., Blaauw, D., A low-voltage processor for sensing applications with picowatt standby mode. *IEEE J. Solid-State Circuits*, 44, 1145–1155, 2009.
41. Fojtik, M., Daeyeon, K., Chen, G., Yu-Shiang, L., Fick, D., Junsun, P., Mingoo, S., Mao-Ter, C., Zhiyoong, F., Blaauw, D., Sylvester, D., A millimeter-scale energy-autonomous sensor system with stacked battery and solar cells. *IEEE J. Solid-State Circuits*, 48, 801–13, 2013.
42. Henry, C.H., Logan, R.A., Merritt, F.R., The effect of surface recombination on current in AlxGa1–xAs heterojunctions. *J. Appl. Phys.*, 49, 3530–3542, 1978.
43. Casey, H.C., Jr., Cho, A.Y., Foy, P.W., Reduction of surface recombination current in GaAs p-n junctions. *Appl. Phys. Lett.*, 34, 594–596, 1979.
44. DeMoulin, P.D., Tobin, S.P., Lundstrom, M.S., Carpenter, M.S., Melloch, M.R., Influence of perimeter recombination on high-efficiency GaAs p/n heteroface solar cells. *IEEE Electron Device Lett.*, 9, 368–370, 1988.

45. Stellwag, T.B., Melloch, M.R., Lundstrom, M.S., Carpenter, M.S., Pierret, R.F., Orientation-dependent perimeter recombination in GaAs diodes. *Appl. Phys. Lett.*, 56, 1658–1660, 1990.
46. Stellwag, T.B., Dodd, P.E., Carpenter, M.S., Lundstrom, M.S., Pierret, R.F., Melloch, M.R., Yablonovitch, E., Gmitter, T.J., Effects of perimeter recombination on GaAs-based solar cells, in: *IEEE Conference on Photovoltaic Specialists*, vol. 1, pp. 442–447, Kissimmee, FL, USA, 1990.
47. Belghachi, A. and Khelifi, S., Modelling of the perimeter recombination effect in GaAs-based micro-solar cell. *Sol. Energy Mater. Sol. Cells*, 90, 1–14, 2006.
48. Chan, D.S.H. and Phang, J.C.H., A method for the direct measurement of solar cell shunt resistance. *IEEE Trans. Electron Devices*, 31, 381–383, 1984.
49. Teran, A.S., Moon, E., Lim, W., Kim, G., Lee, I., Blaauw, D., Phillips, J.D., Energy harvesting for GaAs photovoltaics under low-flux indoor lighting conditions. *IEEE Trans. Electron Devices*, 63, 2820–2825, 2016.
50. Moon, E., Blaauw, D., Phillips, J., Infrared energy harvesting in millimeter-scale GaAs photovoltaics. *IEEE Trans. Electron Devices*, 64, 4554–4560, 2017.
51. Lunt, S.R., Ryba, G.N., Santangelo, P.G., Lewis, N.S., Chemical studies of the passivation of GaAs surface recombination using sulfides and thiols. *J. Appl. Phys.*, 70, 7449–7467, 1991.
52. Kirchner, C., George, M., Stein, B., Parak, W.J., Gaub, H.E., Seitz, M., Corrosion protection and long-term chemical functionalization of gallium arsenide in an aqueous environment. *Adv. Funct. Mater.*, 12, 266–276, 2002.
53. Eisler, C.N., Sheldon, M.T., Atwater, H.A., Enhanced performance of small GaAs solar cells via edge and surface passivation with trioctylphosphine sulfide, in: *2012 38th IEEE Photovoltaic Specialists Conference*, pp. 000937–000940, Austin, TX, 2012.
54. Sheldon, M.T., Eisler, C.N., Atwater, H.A., GaAs passivation with trioctylphosphine sulfide for enhanced solar cell efficiency and durability. *Adv. Energy Mater.*, 2, 339–344, 2012.
55. Lee, I., Lim, W., Teran, A., Phillips, J., Sylvester, D., Blaauw, D., 21.4 A >78%-efficient light harvester over 100-to-100klux with reconfigurable PV-cell network and MPPT circuit, in: *2016 IEEE International Solid-State Circuits Conference (ISSCC)*, pp. 370–371, San Francisco, CA, 2016.
56. Kimovec, R., Helmers, H., Bett, A.W., Topič, M., Comprehensive electrical loss analysis of monolithic interconnected multi-segment laser power converters. *Prog. Photovoltaics Res. Appl.*, 27, 3, 199–209, 2019.

Index

Also Edited by Monika Freunek Müller

Photovoltaic Modeling Handbook

2018 ISBN 9781119363521

Photovoltaic Modeling Handbook provides the reader with a solid understanding of the modeling of photovoltaic devices, from very fundamental theoretic investigations to numerical simulations based on ray tracing and experimental values. The book covers both standard applications, models, new approaches and fields of research such as perovskite materials. Recognized subject-matter experts have written the chapters and they refer to simulation software and the basic literature of the field.

This groundbreaking book guides the reader to their specific application with solid background information in hand and an assessment of which materials might be appropriate. Specifically, the book covers:

- The physics of photovoltaics and the material independent efficiency limits of photovoltaic devices;
- A detailed model of a silicon-based photovoltaic module and a profound introduction to ray-tracing methods for optical numerical models;
- Numerical modeling methods and results pertaining to amorphous silicon;
- The modeling of organic semiconductors as well as an introduction to kinetic Monte Carlo methods to simulate the dynamics in organic semiconductor devices;
- Reviews a few theories on modeling the device physics of chalcogenide thin-film solar cells such as CdTe and Cu(In,Ga)(Se,S)2 (or CIGS) devices;
- Shows the modeling of stacked multi-material solar cells for ultra-high irradiance applications;
- Details the influence of spectral variations both theoretically and experimentally with a special focus on outdoor applications;

- A discussion on indoor applications and shows the resulting ideal choice of materials and the enhanced indoor efficiencies.

Printed in the USA/Agawam, MA
September 16, 2026

910661.009